Mathematical Aspects of Modelling Oscillations and Wake Waves in Plasma

Mathematical Aspects of Modelling Oscillations and Wake Waves in Plasma

E.V. Chizhonkov

CRC Press
Taylor & Francis Group
Boca Raton London New York

CISP

CRC Press is an imprint of the
Taylor & Francis Group, an **informa** business

Translated from Russian by V.E. Riecansky

CRC Press
Taylor & Francis Group
6000 Broken Sound Parkway NW, Suite 300
Boca Raton, FL 33487-2742

First issued in paperback 2021

© 2019 by CISP
CRC Press is an imprint of Taylor & Francis Group, an Informa business

No claim to original U.S. Government works

ISBN 13: 978-1-03-224015-2 (pbk)
ISBN 13: 978-0-367-25527-5 (hbk)

Publisher's Note
The publisher has gone to great lengths to ensure the quality of this reprint but points out that some imperfections in the original copies may be apparent.

Visit the Taylor & Francis Web site at
http://www.taylorandfrancis.com

and the CRC Press Web site at
http://www.crcpress.com

Contents

Preface   ix

Part I: Free plasma oscillations

1.	**Introductory information**	**1**
1.1.	What is breaking?	1
1.2.	Physical model and basic equations	5
1.3.	About initial conditions	13
1.4.	About boundary conditions	16
1.5.	Bibliography and comments	18
2.	**Plane one-dimensional non-relativistic electron oscillations**	**23**
2.1.	Problem statement in Eulerian and Lagrangian variables	23
2.2.	Axial solutions	25
2.3.	'Triangular' solutions	36
2.3.1.	*Simple solutions*	37
2.3.2.	*Composite solutions*	38
2.4.	Numerical–analytical method	40
2.5.	Bibliography and comments	44
3.	**Plane one-dimensional relativistic electron oscillations**	**48**
3.1.	Problem statement in the Eulerian and Lagrangian variables	48
3.2.	Theoretical background of breaking	50
3.2.1.	*Quadratic frequency shift*	51
3.2.2.	*Violation of the property of invariance*	54
3.3.	Method in Lagrangian variables	55
3.4.	Scenario of development and completion of oscillations	57
3.5.	Method in the Eulerian variables	62
3.6.	Artificial boundary conditions	65
3.6.1.	*Full damping of oscillations*	66
3.6.2.	*Linearization of the original equations*	67
3.6.3.	*Accounting for the weak nonlinearity of the original equations*	68
3.6.4.	*Deterioration of the approximation at the boundary*	69
3.7.	Bibliography and comments	71

**4.   Cylindrical one-dimensional relativistic and non-
         relativistic electron oscillations**   **76**
4.1.   Problem statements in Eulerian and Lagrangian variables   76
4.2.   Analytical studies   82
4.2.1.   *Axial solution*   82
4.2.2.   *Perturbation method*   87
4.3.   Finite difference method   89
4.3.1.   *Auxiliary designs. Splitting into physical processes*   89
4.3.2.   *Construction of difference schemes*   91
4.3.3.   *Process scenario*   94
4.5.   Calculation of axial solutions   103
4.5.1.   *Free non-relativistic oscillations*   103
4.5.2.   *Forced relativistic oscillations*   110
4.6.   About spherical oscillations   118
4.6.1.   *Problems formulation*   118
4.6.2.   *Axial solution*   122
4.6.3.   *Perturbation method*   125
4.6.4.   *For numerical modelling*   127
4.7.   Bibliography and comments   129

**5.   Influence of ion dynamics on plane one-dimensional
         oscillations**   **132**
5.1.   Formulation of the problem   132
5.2.   Scaling equations and difference scheme   136
5.3.   Axial solution   140
5.4.   Calculation results   144
5.5.   Bibliography and comments   148

6.   Plane two-dimensional relativistic electron oscillations   **151**
6.1.   Formulation of the problem   151
6.2.   Asymptotic theory   153
6.3.   Difference scheme   157
6.3.1.   *Difference equations in the internal nodes of the grid*   159
6.3.2.   *Implementation of the artificial boundary conditions*   161
6.4.   Numerical experiments   163
6.4.1.   *General remarks*   163
6.4.2.   *Calculations with circular symmetry*   165
6.4.3.   *Quasi-one-dimensional model*   168
6.4.4.   *Small deviation from circular symmetry*   173
6.4.5.   *Significant difference from circular symmetry*   176
6.5.   Bibliography and comments   180

Part II: Plasma wake waves

7.	**Introductory information**	**182**
7.1.	Source equations	182
7.2.	The case of an arbitrary pulse velocity	186
7.2.1.	*Equations in scalar form*	186
7.2.2.	*New coordinates and quasistatics*	187
7.2.3.	*Equations in dimensionless variables*	188
7.2.4.	*Equations in convenient variables*	189
7.3.	The basic formulation of the problem	192
7.3.1.	*Nonlinear statement*	192
7.3.2.	*Linearized formulation*	194
7.4.	'Slow' pulse	196
7.4.1.	*Linearized equations*	196
7.4.2.	*Auxiliary Cauchy problem*	197
7.4.3.	*Numerical–asymptotic method*	201
7.5.	*Bibliography and comments*	203
8.	**Numerical algorithms for the basic problem**	**208**
8.1.	Difference method I	208
8.1.1.	*Construction of a difference scheme*	208
8.1.2.	*Study of schemes in variations*	211
8.1.3.	*The algorithm for implementing the difference scheme I*	214
8.2.	Difference method II	216
8.2.1.	*Construction of a difference scheme*	216
8.2.2.	*Study of schemes in variations*	218
8.2.3.	*Algorithm for the implementation of difference scheme II*	219
8.3.	Difference method III (Linearization method)	221
8.3.1.	*Setting the task in a convenient form*	221
8.3.2.	*Preliminary transformations*	223
8.3.3.	*Difference method III in the linear case*	225
8.3.4.	*Difference method III in the nonlinear case*	226
8.4.	Projection method	228
8.4.1.	*Setting the problem in a convenient form*	228
8.4.2.	*Description of the projection method*	229
8.4.3.	*Numerical implementation of the projection method*	232
8.5.	Numerical experiments and comparison methods	233
8.6.	Bibliography and comments	239
9.	**Additional research**	**246**
9.1.	Axial wake wave solution	246
9.1.1.	*Formulation of the 'truncated' problem*	246
9.1.2.	*Numerical algorithm for solving the 'truncated' problem*	250
9.1.3.	*Calculation results*	251
9.2.	Accounting for the dynamics of ions in the wake wave	257

9.2.1.   *Problem statement in physical variables*   257
9.2.2.   *Statement of the problem in convenient variables*   260
9.2.3.   *Solution method*   262
9.2.4.   *Calculation results*   266
9.3.   Elliptical pulse   268
9.3.1.   *Formulation of the problem*   268
9.3.2.   *Difference scheme and solution method*   273
9.3.3.   *Calculation results*   277
9.4.   Bibliography and comments   279

Conclusion   281
References   284
Index   291

Preface

This book is fully devoted to one of the most popular areas in modern mathematics – numerical modelling. However, its appearance is due to a qualitative, one might say revolutionary, change in the situation in another field of science – in physics. The fact is that at present there is an urgent need to rethink ideas about the interaction of electromagnetic fields with matter. Let us give a weighty argument in favour of this statement from the popular science article 'Why do we need high-intensity laser pulses?' [43] written by L.M. Gorbunov, an outstanding specialist in the field of plasma physics (1934–2007).

The simplest atom is the hydrogen atom. In it, a single electron moves around the nucleus (proton). The electric field, due to which the two particles are held near each other, is about $5 \cdot 10^9$ V/cm. This is a very strong field compared to those found in everyday life that surrounds us. For example, the breakdown of such a good insulator, like mica, occurs at $2 \cdot 10^6$ V/cm (the field here is one thousand times weaker!). The classical theory, in essence, is a 'theory of small perturbations', that is, it assumes a small external field compared to fields that keep the atomic systems in equilibrium. But in recent years lasers have been created that generate high-intensity ultra-short light pulses. During experiments with them, new physical phenomena were discovered, the possibilities of using pulses in various fields, ranging from nuclear physics and astrophysics to medicine, are analyzed. These pulses have a duration of less than 1 picosecond (that is, less than 10^{-12} s). Their length in space is less than 300 microns, which is less than a third of a millimeter. The wavelength of radiation is usually about 1 micron, and it belongs to the infrared range. Tens or hundreds of wavelengths are stacked on the pulse length. The energy that carries such an pulse can reach hundreds of joules, and power – up to 10^{15} watts. This value is called the petawatt. It far exceeds the total capacity of all power plants in the world. If such a pulse is focused on a pad with a radius of 10 μm, the radiation intensity will reach $3 \cdot 10^{20}$ W/cm^2, and the electric field strength will be about 10^{12} V/cm. In other words, such fields are thousands of times stronger than the fields inside a hydrogen atom. Of course, these numbers explain well why, in many countries, the propagation of such pulses and their interaction with matter is being conducted on a broad front: it is about building a new physical theory that describes strongly nonlinear effects.

A good example of a new theory is the so-called wake wave in a plasma. This term originated from the analogy with the movement of a ship on water. When propagating in a plasma, a short laser pulse excites a wake plasma wave behind it, which is a two-dimensional charge density wave that propagates in the direction of the pulse and is bounded in the transverse direction. The theoretically predicted effect of the excitation of a wake plasma wave by a laser pulse was further confirmed in a number of experiments, where the spatial structure of the wake fields was studied. An important feature of a two-dimensional nonlinear wake wave, which significantly distinguishes it from a one-dimensional charge density wave, is that its characteristics change with distance from the laser pulse. a noticeable curvature of the wave front of the wake fields occurs at a distance from the back front of the pulse. In addition, at a certain distance behind the pulse, the wake wave breaks, transferring its energy to the plasma particles. This leads to uncontrolled capture and acceleration of the electrons in the area of wake wave breaking. At the same time, the wake wave breaking is accompanied by intense short-wave electromagnetic radiation, also recorded experimentally.

Currently, the wake waves are primarily used to produce bunches of high-energy electrons. In a wake-wave laser accelerator, the electrons can be accelerated to high energies at significantly shorter distances than in traditional accelerators. Recently, the possibility of using plasma wake waves for the generation of electromagnetic radiation in the terahertz frequency range is also being considered. Considerable interest in the generation and registration of terahertz waves is associated with the wide possibilities of its use in medicine, biology, astrophysics, safety and control systems, as well as in other important areas of science and technology. Therefore, it seems necessary to study in detail the spatial structure of wake waves to obtain bunches of accelerated particles and pulses of terahertz radiation with the required parameters. In addition, the study of the space-time structure of wake fields is of considerable interest also from a purely academic point of view, namely, for the further development of the theory of nonlinear waves and gradient catastrophes that arise when they break.

The main tool for numerical studies of wake waves in plasma are the so-called PiC models (Particle-in-Cell). The program codes implementing them are well known: OSIRIS [134], QUICKPIC [140], VORPAL [155], VLPL [158], OOPIC [168], etc. The modern presentation of this approach is thoroughly presented in the monograph by Yu.N. Grigor'ev, V.A. Vshivkov and M.P. Fedoruk 'Numerical modelling by particle-in-cell methods' [50]. The use of the PiC models is of a pronounced imitative nature, which is very attractive for researchers-physicists, however, it is based on the use of supercomputers. And this requires the simultaneous and coordinated work of a large number of specialists in various fields, that is, it is very expensive. In addition, the results of numerical simulation still have to be interpreted sooner or later in terms of differential equations, ordinary or with partial derivatives, which is associated with the traditional description (representation) of various effects in theoretical physics.

Given these factors, in the Lebedev Physical Institute of the Russian Academy of Science under the guidance of Doctor of physical and mathematical Sciences, Chief researcher, Laureate of the State prize of the USSR Prof. L.M. Gorbunov a hydrodynamic model of a wake wave exited by a laser pulse was developed. It should be noted that the hydrodynamic models have a very limited scope for modelling in plasma compared to the particle models. But hydrodynamics allows a more detailed study of the effects observed in the 'zone of responsibility' of differential equations, and there is no need to 'throw an extra bridge' from numerical experiments to theoretical conclusions, since rthe same objected is analyzed analytically, asymptotically and numerically this case..

The fundamental difference between the PiC models and the hydrodynamic model should be emphasized: the former are based on a kinetic equation of the Vlasov type [38] or its characteristics [139], and the hydrodynamic model is based on differential equations of a continuous medium. Moreover, there is no reason to assert that the hydrodynamic model under consideration is a consequence of the formal limiting transition or averaging of some PiC model. This requires additional and very deep research. The hydrodynamic model is a meaningful addition to the PiC models, that is, a qualitatively different way of reflecting the phenomenon observed in nature – the wake wave excited in a plasma by a short high-intensity laser impulse. Without limiting generality, we can say that all content of the book is closely connected with the study of the properties of this hydrodynamic model and its modifications, that fully meets the 'tastes' of the researcher in the field of computational mathematics.

The construction of the model studied in the book was preceded by a series of studies on various types of non-linearities in plasma (ponderomotive, thermal, ionization, etc.), closely related to the known self-focusing effect of laser beams [73, 105]. In these works, the author for the purpose of numerical simulation designed special algorithms based on kinetically consistent difference schemes for the Eulerian equations [55, 56, 97], methods for calculating thermal waves [84] and symmetric implicit schemes for the nonlinear Schrödinger equation [21, 30, 63]. The results obtained are of independent interest in the calculations of laser–plasma interactions, are reflected in bibliography and commentaries, but are not placed in the book due to their somewhat different scientific focus.

The basis of the hydrodynamic model of the wake wave is a differential–algebraic system of nonlinear equations, which is very difficult for analytical and asymptotic research methods. Therefore, numerical algorithms come to the fore as an analysis tool. On the other hand, having no idea about the qualitative properties of the solution, it is very difficult to construct an effective numerical method. Moreover, a special role in the dynamics of the wake wave belongs to the breaking effect. This means that in such problem statements a combination of different approaches is required.

The constructed model proved to be very informative. In particular, for the first time, an unexpected result was obtained. It was found that a three-dimensional nonlinear wake wave, excited by an axially symmetric laser pulse, can break not on the axis of symmetry of the system, but outside this axis. For a detailed study of this effect, it was necessary to involve simpler models – electron oscillation models, with the help of which it was possible to analyze in detail the effect of off-axis breaking both numerically and analytically. In other words, the study of the wake wave led to a rethinking and refinement of the well-known results associated with large-amplitude electron oscillations in a cold plasma [131].

Another new and unusual result, established in the framework of the simplification of the model under study, is the instability of the shape of the electron density function at the moment of breaking of plasma oscillations. If the initial perturbation is axially symmetric, then the breaking (that is, the singularity of the electron density function) arises simultaneously on the circle of a certain nonzero radius. However, if the cross-sections of the initial electron density perturbation are ellipses, then the singularity also occurs simultaneously, but only at two points. Moreover, a qualitative change in the final picture of breaking does not depend on the quantitative deviation of the ellipse from the circle: any arbitrarily small deviation of the ratio of the ellipse semiaxes from unity entails the conversion to infinity of the electron density at only two points.

It should be noted that the hydrodynamic model of the wake wave is, in a sense, open. It can be refined both by adding physical factors (for example, ion dynamics) and by complicating the mathematical formulation (for example, taking into account plasma inhomogeneities in the form of given functions or complicating equations by introducing small parameters with additional terms). Such modifications can greatly change the requirements for the corresponding algorithms of numerical simulation. In the process of developing new numerical schemes (methods), it was often necessary to resort to consultations of the most qualified specialists in the field of numerical analysis and solving applied problems. In this regard, Academician of the Russian Academy of Sciences N.S. Bakhvalov (1934–2005) and Prof. V.I. Lebedev (1930–2010) deserve special thanks for the attention to the work and good memory.

Let us determine the place of the present monograph in the series of modern scientific literature. At the beginning of the century, the fundamental scientific work 'Mathematical problems of the numerical solution of hyperbolic systems of equations' by A.G. Kulikovsky, N.V. Pogorelov and A.Yu. Semenov [69] was published. It provides a detailed description of various mathematical aspects of the numerical solution of hyperbolic systems of partial differential equations. At the same time, all the content is presented in close connection with such important mechanical applications of these systems as gas dynamics, shallow water theory, magnetic hydrodynamics and the mechanics of deformable solids. New times bring new challenges. In this regard, the proposed book should in the first place be considered as an addition to the specified work or an extension of the material set forth in it. More specifically, the theme of modelling oscillations and wake waves excited by short high-intensity laser pulses in a plasma is elaborated in detail in this book. One should also note, as a 'closest neighbour on the bookshelf' a very recent monograph by a famous specialist in the field of computational plasma physics K.V. Brushlinsky 'Mathematical and computational problems of magnetic gas dynamics' [31]. It very carefully and instructively discusses both the properties of MHD equations and the ideas and methods of their numerical analysis. This monograph deals with the modelling of completely different (one might say, 'classical') effects of dynamics and statics in a plasma; the material is presented, first of all, from the standpoint of a physicist-calculator. In this regard, the proposed book is a useful addition to this monograph, both in terms of other problem statements, and in terms of the form for presenting the results.

The book consists of a preface, two parts, divided into nine chapters, and conclusion. Each chapter is provided with a brief description of its content, contains several paragraphs and ends with a bibliography and comments on the presented topic. In order to simplify the presentation of the material, the first six chapters are devoted to various aspects of plasma oscillation modelling, and the next three to the wake waves.

The book presents the results of about forty scientific articles published in domestic and foreign, peer-reviewed journals; they have been repeatedly reported at international and Russian scientific conferences in mathematics and physics. Ongoing research during 1993–2007, 2009–2011. received grant support from the Russian Foundation for Basic Research and other organizations (project managers: in mathematics – N.S. Bakhvalov and A.A. Kornev, in physics – L.M. Gorbunov). Multidimensional calculations, due to the high computational costs, have been performed mainly on the SKIF MSU Chebyshev supercomputer.

As a rule, traditional scientific monographs sum up the results of scientific research in a certain subject area, fixing the achieved level of knowledge or the point of view of the authors. In this case, the situation is somewhat different. The material presented in the book does not claim any completeness at all. On the contrary: the main focus of the presentation is an orientation towards the future, that is, drawing the attention of researchers to new problem statements, studying the properties of their solutions and constructing various methods for analyzing them. Therefore, it is assumed that the main readers of the book will be young scientists, postgraduates and students, as well as engineers and researchers in applied fields of knowledge. However, it seems plausible that even established specialists, both in the field of differential equations and in the field of computational mathematics, will be able to find here some unexpected and interesting information. Some of the possible directions for further research are reflected in the conclusion.

The author is deeply grateful to N.E. Andreev and A.A. Frolov, joint activities with which allowed the author to develop a mathematical apparatus for the study of plasma oscillations and wake waves, as well as S.V. Milyutin – for qualified assistance in the field of parallel computing in numerical simulation. An exceptional gratitude, combined with a good memory, is addressed to L.M. Gorbunov, whose personality in the formulation of specific tasks and the discussion of methods for solving them, as well as in the formation of a completely different – physical – view of the world around us can not be overestimated.

Part I

FREE PLASMA OSCILLATIONS

1

Introductory information

The chapter introduces the concept of breaking – the effect arising from the study of the hydrodynamic model of noninteracting particles. In addition, the basic system of equations characterizing the dynamics of a collisionless plasma is determined and the simplest one-dimensional plane subsystems are derived from it: relativistic and non-relativistic. In order to become familiar with the mathematical aspects of the breaking effect for subsystems, initial and boundary conditions are discussed, that is, basic problem statements are formulated.

1.1. What is breaking?

Let us get acquainted with the concept of breaking arising in the study of the movements of a continuous medium in the framework of the hydrodynamic model of noninteracting particles. We carry out the presentation, following [58].

We use the Lagrange construction: the motion of a substance can be described by setting the instantaneous position $\mathbf{r}$ of each particle (i.e., its Eulerian coordinates x, y, z) depending on its initial position ψ (i.e. its Lagrangian coordinates ξ, η, ζ) and time t:

$$\mathbf{r} = \mathbf{r}(\psi, t).$$

It is convenient to imagine that the motion consists in a continuous transformation of the vector field **r** depending on the continuously varying scalar parameter t. The law of this transformation is determined by setting the velocity field **u**:

$$\frac{d\mathbf{r}}{dt} = \mathbf{u}.$$

In turn, the velocity field can be defined by one or another law depending on the operating forces, but this is not important for the essence of the matter.

Let us consider the simplest case when no external forces act on the substance. Moreover, we will lower also internal forces, i.e., any interaction of particles of a substance among themselves. In other words, in the equations we neglect stresses, or, in the case of a liquid or gas, we neglect pressure and viscosity. We will assume that each particle of matter moves at a constant speed, i.e. there is a uniform and rectilinear movement of individual particles:

$$\mathbf{r} = \psi + (t - t_0)\mathbf{u},$$

and for each particle the value of **u** is constant. This means that $\mathbf{u} = \mathbf{u}(\psi)$, but does not depend on time. Even in this simple case, consideration of a continuous medium, that is, not one particle, but a whole ensemble of particles, leads to interesting and non-trivial results.

For definiteness, we assume that the initial density of the medium at the time $t_0 = 0$ is constant everywhere: $n = n_0$. Motion is given in the Lagrangian system by the simple formula

$$\mathbf{r} = \psi + t\mathbf{u}(\psi).$$

Let us study the question of how the density distribution changes over time.

The specificity of the problem is that the **trajectories of particles** that do not interact with each other **can intersect**. To denote this situation, we will use the term '**breaking**'.

Take the simplest case of one-dimensional motion

$$x = \xi + tv(\xi),$$

which is considered in detail in the book [59]. We note a useful corollary of the formula

$$\frac{dx}{d\xi} = 1 + t\frac{dv}{d\xi}.$$

Here, the variable t is understood as a parameter, so the symbol of the ordinary derivative is used, and not the partial one.

Two particles, whose initial positions were ξ_1 and ξ_2, will be at the same point x', if the equality

$$x' = x_1 = \xi_1 + t'v(\xi_1) = \xi_2 + t'v(\xi_2) = x_2$$

holds and at the moment of time

$$t' = \frac{\xi_2 - \xi_1}{v(\xi_1) - v(\xi_2)}.$$

Such a moment exists in the future, that is, for $t > 0$, if for $\xi_2 > \xi_1$ we have $v(\xi_2) < v(\xi_1)$.

For two neighbouring particles

$$\xi_2 = \xi_1 + d\xi, \quad t' = -\left(\frac{dv}{d\xi}\right)^{-1}.$$

The amount of substance enclosed in the small interval $d\xi$ is $dm = n_0 d\xi$. The value of dm is conserved in the process of movement. We can write $dm = ndx$, where dx must be taken in accordance with the choice of $d\xi$. Thus, we get

$$n = \frac{dm}{dx} = \frac{dm}{d\xi}\cdot\frac{d\xi}{dx} = n_0\left(\frac{dx}{d\xi}\right)^{-1},$$

which is equivalent to

$$n = n_0\left(1 + t\frac{dv}{d\xi}\right)^{-1}.$$

From this it follows that the **moment of intersection of neighbouring trajectories** t' is precisely **the moment when the density of the substance becomes infinite**. The density versus x-coordinate curves with infinite peaks are analyzed in detail in [59].

Thus, the fundamentally important concept of breaking within the hydrodynamic model of noninteracting particles is this: with the Lagrangian description of the medium the trajectories two or more particles intersect; in the Euler's description of the medium, this time point and this point in space correspond to the infinity of the density function.

Let us now consider an interesting case for the further presentation, when particles oscillate around some of their equilibrium positions, i.e., their trajectories are described by the formula

$$x = \xi - A(\xi)\cos[\omega(\xi)t], \quad \omega(\xi) = 1 + \sigma A^2(\xi). \tag{1.1}$$

In this formula, the initial displacement of particles relative to the equilibrium position is given by the function $A(\xi)$, and the shift of the fundamental frequency of oscillations equal to one is quadratically dependent on the initial displacement when the value of the parameter $\sigma \neq 0$. A little running ahead, we note that such trajectories at $A(\xi) = A_0\xi\exp\{-2\xi^2 / \rho_*^2\}$, ρ_*, $A_0 = \text{const}$ are typical of nonlinear electronic oscillations in a plasma; the case $\sigma = 0$ corresponds to linear oscillations.

Suppose that at the point x at the moment of time t, an intersection of the trajectories of neighbouring particles, characterized by the Lagrangian coordinates ξ and $\xi + \Delta\xi$, occurred. This means that, in addition to equalities (1.1), the following relation holds:

$$x = \xi + \Delta\xi - A(\xi + \Delta\xi)\cos[\omega(\xi + \Delta\xi)t]. \tag{1.2}$$

We subtract from (1.2) the first equality of (1.1), add the expression $\pm A(\xi + \Delta\xi)\cos[\omega(\xi)t)]$ in the right-hand side, group the terms and divide by term by $\Delta\xi$. We will have

$$0 = 1 - \frac{A(\xi + \Delta\xi) - A(\xi)}{\Delta\xi}\cos[\omega(\xi)t)] -$$
$$A(\xi + \Delta\xi)\frac{\cos[\omega(\xi + \Delta\xi)t)] - \cos[\omega(\xi)t)]}{\Delta\xi}. \tag{1.3}$$

Passing to (1.3) to the limit as $\Delta\xi \to 0$, we obtain the equation $\frac{\partial x}{\partial \xi} = 0$ or, which is the same,

$$0 = 1 - A'(\xi)\cos[\omega(\xi)t)] + A(\xi)\sin[\omega(\xi)t)]\omega'(\xi)t.$$

After the introduction of the auxiliary angle φ in the usual way, the resulting equation is converted to

$$A'(\xi)\sqrt{1 + 4t^2\sigma^2 A^4(\xi)}\cos[\omega(\xi)t) + \varphi] = 1. \tag{1.4}$$

The possibility of breaking is determined solely by the solvability of equation (1.4). Since the factor in the cos function monotonously increases (in modulus) as t increases and all functions in (1.4) are continuous, an explicit formula of the form $A(\xi) = A_0\xi\exp\{-2\xi^2 / \rho_*^2\}$ guarantees the occurrence of an oscillation breaking effect.

With a sufficiently small amplitude of oscillations $A(\xi) = O(\varepsilon)$, it is easy to estimate the asymptotic behaviour with respect to the time of start of the breaking moment. Putting $|A'(\xi)|\sqrt{1 + 4t^2\sigma^2 A^4(\xi)} = O(1)$ and $|A'(\xi)| = O(\varepsilon)$, we have

$$t = O\left(\frac{1}{|A'(\xi)| A^2(\xi)}\right) = O(\varepsilon^{-3}).$$

Note that the breaking effect in this situation necessarily arises due to the dependence of the oscillation frequency on the amplitude, that is, on the condition $\omega'(\xi) \neq 0$. In the case where there is no frequency shift [102], small amplitude oscillations can continue indefinitely long.

1.2. Physical model and basic equations

Let us define plasma as a mixture of cold, ideal, relativistic electron and non-relativistic ionic liquids, then the system of hydrodynamic equations describing it together with Maxwell's equations in vector form will look like:

$$\frac{\partial n_e}{\partial t} + \text{div}(n_e \mathbf{v}_e) = 0, \tag{1.5}$$

$$\frac{\partial \mathbf{p}_e}{\partial t} + (\mathbf{v}_e \cdot \nabla)\mathbf{p}_e = e\left(\mathbf{E} + \frac{1}{c}[\mathbf{v}_e \times \mathbf{B}]\right), \tag{1.6}$$

$$\gamma = \sqrt{1 + \frac{|\mathbf{p}_e|^2}{m_e^2 c^2}}, \tag{1.7}$$

$$\mathbf{v}_e = \frac{\mathbf{p}_e}{m_e \gamma}, \tag{1.8}$$

$$\frac{\partial n_i}{\partial t} + \text{div}(n_i \mathbf{v}_i) = 0, \tag{1.9}$$

$$\frac{\partial \mathbf{v}_i}{\partial t} + (\mathbf{v}_i \cdot \nabla)\mathbf{v}_i = \frac{e_i}{m_i}\left(\mathbf{E} + \frac{1}{c}[\mathbf{v}_i \times \mathbf{B}]\right), \tag{1.10}$$

$$\frac{1}{c}\frac{\partial \mathbf{E}}{\partial t} = -\frac{4\pi}{c}\left(en_e \mathbf{v}_e + e_i n_i \mathbf{v}_i\right) + \text{curl } \mathbf{B}, \tag{1.11}$$

$$\frac{1}{c}\frac{\partial \mathbf{B}}{\partial t} = -\text{curl } \mathbf{E}, \tag{1.12}$$

where e, e_i, m_e, m_i are the charges and masses of electrons and ions, respectively (here the electron charge has a negative sign: $e < 0$); c is the speed of light; n_e, $\mathbf{p}_e$, $\mathbf{v}_e$ – concentration, momentum and velocity of electrons; n_i, v_i is the concentration and velocity of ions; γ is the Lorentzian factor; $\mathbf{E}$, $\mathbf{B}$ – vectors of electric and magnetic fields.

The system of equations (1.5)–(1.12) is one of the simplest models of a collisionless plasma for which the so-called quasi-hydrodynamic description given above and called simply hydrodynamic in the literature is valid. This approximation is often called the *equations of two-fluid magnetic hydrodynamics of a 'cold' plasma*; it is well known and described in sufficient detail in textbooks and monographs on plasma physics (see, for example, [6, 40, 85, 86]).

Of interest is the behaviour of the solution of the Cauchy problem for the system (1.5)–(1.12), therefore, we assume that for $t = 0$, suitable initial conditions are given:

$$\mathbf{p}=\mathbf{p}^0, \quad n=n^0, \quad \mathbf{E}=\mathbf{E}^0, \quad \mathbf{B}=\mathbf{B}^0. \tag{1.13}$$

The term 'suitable' in this case means that they do not contradict Maxwell's equations and / or their consequences, for example,

$$\mathrm{div}\mathbf{B}^0 = 0, \quad \mathrm{div}\mathbf{E}^0 = 4\pi(n^0 - n_i^0),$$

where n_i^0 is the initial distribution of the ion density in the plasma.

If the plasma motion is considered in the whole space $\mathbb{R}^3$ and the initial conditions have some specificity, then equation (1.6) can be transformed to a more convenient form. Here it is meant that the desired solution in a particular case can be described by simpler relations.

Statement 1.2.1. *Let the Cauchy problem for equations (1.5)– (1.12) have a sufficiently smooth (twice continuously differentiable with respect to all independent variables and bounded together with the indicated derivatives) solution with the necessary asymptotic decay at infinity and initial functions* $\mathbf{B}^0(\mathbf{x}, t)$, $\mathbf{p}_e{}^0(\mathbf{x}, t)$ *are related by*

$$\mathbf{B}^0(\mathbf{x},t)+\frac{c}{e}\mathrm{curl}\,\mathbf{p}_e{}^0(\mathbf{x},t)=0. \tag{1.14}$$

Then equation (1.6) can be represented in the following form:

$$\frac{\partial \mathbf{p}_e}{\partial t}=e\mathbf{E}-m_e c^2 \nabla\gamma. \tag{1.15}$$

Proof. If the functions $\mathbf{v}_e$, $\mathbf{p}_e$, γ are connected by algebraic relations (1.7), (1.8)

$$\mathbf{v}_e=\frac{\mathbf{p}_e}{m_e\gamma}, \quad \gamma=\sqrt{1+\frac{|\mathbf{p}_e|^2}{m_e^2 c^2}},$$

then the equity test is established by direct verification

$$(\mathbf{v}_e\cdot\nabla)\mathbf{p}_e = m_e c^2 \nabla\gamma -[\mathbf{v}_e\times\mathrm{curl}\,\mathbf{p}_e]. \tag{1.16}$$

Here the operator ∇ applied to the scalar function γ has the usual meaning of the operator grad.

After substituting (1.16) into equation (1.6) we get

$$\frac{\partial \mathbf{p}_e}{\partial t} = e\mathbf{E} - m_e c^2 \nabla \gamma + \frac{e}{c}[\mathbf{v}_e \times \mathbf{A}], \tag{1.17}$$

where notation for auxiliary vector field is used

$$\mathbf{A} = \mathbf{B} + \frac{c}{e}\operatorname{curl} \mathbf{p}_e. \tag{1.18}$$

We show that, under condition (1.14), the field **A** vanishes identically at an arbitrary time instant. Apply the rot operation to the vector equation (1.17) and eliminate curl **E** from the obtained relation, using the Maxwell equation (1.12). As a result, we obtain the equation

$$\frac{\partial \mathbf{A}}{\partial t} = \operatorname{curl}[\mathbf{v}_e \times \mathbf{A}].$$

In addition, the equality div **A** = 0 is valid, which is a consequence of the relations (1.12), (1.14) and (1.18). This makes it possible to consider for an auxiliary vector field **A** a problem with the initial condition (1.14):

$$\frac{\partial \mathbf{A}}{\partial t} = \operatorname{curl}[\mathbf{v}_e \times \mathbf{A}], \quad \operatorname{div} \mathbf{A} = 0,$$

$$\mathbf{A}\big|_{t=0} = \mathbf{0}. \tag{1.19}$$

We multiply the non-stationary equation in (1.19) scalarly by the vector function **A** and integrate over the whole space $\mathbb{R}^3$. Will get

$$\frac{1}{2}\frac{\partial \|\mathbf{A}\|^2}{\partial t} = \int_{\mathbb{R}^3} (\operatorname{curl}[\mathbf{v}_e \times \mathbf{A}], \mathbf{A})\, d\mathbf{x}, \quad \text{where} \quad \|\mathbf{A}\|^2 = \int_{\mathbb{R}^3}\left(\sum_{i=1}^{3}|A_i(\mathbf{x},t)|^2\right) d\mathbf{x}. \tag{1.20}$$

We transform the integral on the right side of equation (1.20). Note first that theh following equality is true

$$\operatorname{curl}[\mathbf{v}_e \times \mathbf{A}] = (\mathbf{A}\cdot\nabla)\mathbf{v}_e - (\mathbf{v}_e\cdot\nabla)\mathbf{A} - \mathbf{A}\operatorname{div}\mathbf{v}_e + \mathbf{v}_e\operatorname{div}\mathbf{A}.$$

Taking into account the solenoidal form of the vector field **A**, in the resulting equality we leave only three terms. Consider the integral with the last two of the remaining terms:

$$I_1 = - \int_{\mathbb{R}^3} \left((\mathbf{v}_e \cdot \nabla) \mathbf{A} + \mathbf{A} \operatorname{div} \mathbf{v}_e, \mathbf{A} \right) dx.$$

After integrating by parts in it, the first term will have

$$I_1 = -\frac{1}{2} \int_{\mathbb{R}^3} \left(\mathbf{A} \operatorname{div} \mathbf{v}_e, \mathbf{A} \right) dx \leqslant \frac{3}{2} S(t) \| \mathbf{A} \|^2,$$

where $S(t) = \max\limits_{x \in \mathbb{R}^3} \max\limits_{1 \leqslant i, j \leqslant 3} \left| \dfrac{\partial v_i}{\partial x_j} \right|.$

Here $v_i (i = 1, 2, 3)$ denotes the components of the vector $\mathbf{v}_e(\mathbf{x}, t)$. The following integral is valid for the remaining integral.

$$I_2 = \int_{\mathbb{R}^3} \left((\mathbf{A} \cdot \nabla) \mathbf{v}_e, \mathbf{A} \right) dx \leqslant 3 S(t) \| \mathbf{A} \|^2 .$$

We note that the necessary asymptotic behaviour of decreasing functions at infinity is required exclusively for the finiteness of the integrals being evaluated and the integration by parts [5], [88].

Using the obtained estimates of the integrals in relation (1.20), we arrive at the desired expression

$$\frac{\partial \| \mathbf{A} \|^2}{\partial t} \leqslant 9 S(t) \| \mathbf{A} \|^2,$$

which, together with the initial condition (1.14) $\|\mathbf{A}\| = 0$, guarantees the triviality of the vector field $\mathbf{A}$ at an arbitrary time based on Gronwall inequality in differential form [64]. The statement is proven.

For definiteness, the transformation of equation (1.6) to the simpler form (1.15) is carried out in the case of the space $\mathbb{R}^3$. It is easy to generalize it to the case of $\mathbb{R}^q$, $q = 1, 2$.

When modelling plasma oscillations, it should be borne in mind that it is very difficult to establish the existence of the breaking effect, since it can be deeply hidden 'in the depths' of the model under consideration (1.5)–(1.12). Therefore, it is reasonable to simplify the equations to such an extent that the boundary between the presence of a breaking effect and the absence of it is found.

We introduce in the space $\mathbb{R}^3$ the usual rectangular coordinate system $OXYZ$ and apply the assumptions that

- ions, due to the multiple excess of electrons by mass, are considered immobile;
- the solution is determined only by the x-components of the vector functions $\mathbf{p}_e$, $\mathbf{v}_e$, $\mathbf{E}$;
- there is no dependence in these functions on the variables y and z, that is, $\partial/\partial y = \partial/\partial z = 0$.

Then, from the system (1.5)–(1.12) the non-trivial equations follow:

$$\frac{\partial n}{\partial t} + \frac{\partial}{\partial x}(nv_x) = 0, \quad \frac{\partial p_x}{\partial t} + v_x \frac{\partial p_x}{\partial x} = eE_x,$$

$$\gamma = \sqrt{1 + \frac{p_x^2}{m^2 c^2}}, \quad v_x = \frac{p_x}{m\gamma}, \quad \frac{\partial E_x}{\partial t} = -4\pi e n v_x. \tag{1.21}$$

In this situation, the subscript e, which characterizes the connection of variables with the electron component of the plasma, is of low content and is therefore omitted.

We introduce dimensionless quantities

$$\rho = k_p x, \quad \theta = \omega_p t, \quad V = \frac{v_x}{c}, \quad P = \frac{p_x}{mc}, \quad E = -\frac{eE_x}{mc\omega_p}, \quad N = \frac{n}{n_0},$$

where $\omega_p = (4\pi e^2 n_0/m)^{1/2}$ is the plasma frequency, n_0 is the value of the unperturbed electron density, $k_p = \omega_p/c$. In the new variables, the system (1.21) takes the form

$$\frac{\partial N}{\partial \theta} + \frac{\partial}{\partial \rho}(NV) = 0, \quad \frac{\partial P}{\partial \theta} + E + V\frac{\partial P}{\partial \rho} = 0,$$

$$\gamma = \sqrt{1 + P^2}, \quad V = \frac{P}{\gamma}, \quad \frac{\partial E}{\partial \theta} = NV. \tag{1.22}$$

From the first and last equations (1.22) it follows

$$\frac{\partial}{\partial \theta}\left(N + \frac{\partial E}{\partial \rho}\right) = 0.$$

This relationship is true both in the absence of plasma oscillations ($N \equiv 1$, $E \equiv 0$), and in their presence, therefore we have a simpler expression for the electron density:

$$N = 1 - \frac{\partial E}{\partial \rho}. \tag{1.23}$$

Now, excluding from the system (1.22) the density N and the factor γ, we arrive at the equations describing the free plane one-dimensional relativistic oscillatory motion of electrons in a cold ideal plasma:

$$\frac{\partial P}{\partial \theta} + E + V\frac{\partial P}{\partial \rho} = 0, \quad \frac{\partial E}{\partial \theta} - V + V\frac{\partial E}{\partial \rho} = 0, \quad V = \frac{P}{\sqrt{1+P^2}}. \tag{1.24}$$

Let us introduce the abbreviation P1RE (Plane 1-dimension Relativistic Electron oscillations) for this system. It seems that the system P1RE (1.24) is one of the most simple (if not the simplest), the solutions of which have the effect of breaking.

Finally, we will take the last step in simplifying the original system (1.5)–(1.12): 'isolating' equation (1.24) to such an extent that the desired breaking effect as a property of the solution disappears. We make an additional assumption that the electron velocity is assumed to be essentially nonrelativistic, i.e.

$$P \approx V, \quad \frac{\partial P}{\partial \rho} \approx \frac{\partial V}{\partial \rho}, \quad \frac{\partial P}{\partial \theta} \approx \frac{\partial V}{\partial \theta}.$$

In this case, we arrive at the equations describing the free plane one-dimensional non-relativistic oscillatory movements of electrons in a cold ideal plasma:

$$\frac{\partial V}{\partial \theta} + E + V\frac{\partial V}{\partial \rho} = 0, \quad \frac{\partial E}{\partial \theta} - V + V\frac{\partial E}{\partial \rho} = 0. \tag{1.25}$$

The solutions of this system, we denote it as P1NE (Plane 1-dimension Nonrelativistic Electron oscillations), no longer have the desired breaking effect. Nevertheless, the equations (1.25) are not trivial, and a certain attention will be paid to the study of their properties.

We clarify that the immediate interest is not instantaneous breaking (during the first period of oscillation), but the effect that occurs after a certain number of periods, i.e., a consequence of the quasiperiodic nature of the trajectories.

Since the effect of the breaking of oscillations in the Eulerian and Lagrangian description of a medium manifests itself in different ways, it will be very useful to operate with the equations not only in the Eulerian but also in Lagrangian variables. We give them.

Let's start with an analogue of equations (1.25). Let some particle of the medium be characterized by the Lagrangian variable ρ^L, then from (1.25) we have the equations describing its dynamics:

$$\frac{dV(\rho^L,\theta)}{d\theta} = -E(\rho^L,\theta), \quad \frac{dE(\rho^L,\theta)}{d\theta} = V(\rho^L,\theta), \qquad (1.26)$$

where $d/d\theta = \partial/\partial\theta + V\partial/\partial\rho$ is the total time derivative.

Recall that the quantity R, which determines the displacement of a particle with the Lagrangian variable ρ^L, i.e., the trajectory

$$\rho(\rho^L,\theta) = \rho^L + R(\rho^L,\theta),$$

satisfies the equation

$$\frac{dR(\rho^L,\theta)}{d\theta} = V(\rho^L,\theta).$$

From this it follows that the values of $R(\rho^L, \theta)$ and $E(\rho^L, \theta)$ coincide up to a constant, which is calculated from the condition that the electric field is zero in the absence of displacements. In other words, in the case of flat one-dimensional oscillations, the relation $R(\rho^L, \theta) \equiv E(\rho, \theta)$ is valid, and the set of equations (1.26) in Lagrangian variables takes the form

$$\frac{dV}{d\theta} = -R, \quad \frac{dR}{d\theta} = V. \qquad (1.27)$$

Acting in a similar way, we obtain from (1.24) the equations of motion of a particle of a medium with allowance for relativism:

$$\frac{dP}{d\theta} = -R, \quad \frac{dR}{d\theta} = V \equiv \frac{P}{\sqrt{1+P^2}}. \tag{1.28}$$

Note that the equations describing the dynamics of electrons in the Eulerian variables (1.24), (1.25) differ from the equations in the Lagrangian variables (1.27), (1.28) in principle: the new equations are ordinary differential equations. In them, the desired functions (momentum–velocity and displacement) depend only on time, and the Lagrangian variable ρ^L, which characterizes a single particle, participates in the form of a parameter, that is, only the initial data can depend on it. This fact emphasizes that in this model we are talking about particles that do not interact with each other: their dynamics are individualized by the initial conditions, and the trajectories are likely to intersect.

1.3. About initial conditions

Consider the situation when electrons in a plasma are derived from an equilibrium position due to the propagation of a short laser pulse. According to the kinetic model of the propagation of laser pulses in a plasma [153] (see also [10]), equation (1.6) should be modified taking into account the slowly varying complex amplitude of the high-frequency laser field $a(\mathbf{x}, t)$, $\mathbf{x} \in \mathbb{R}^q$, $q = 1, 2, 3$ (the so-called *envelope*) as follows:

$$\frac{\partial \mathbf{p}_e}{\partial t} + \left(\mathbf{v}_e \nabla\right)\mathbf{p}_e = e\left(\mathbf{E} + \frac{1}{c}\left[\mathbf{v}_e \times \mathbf{B}\right]\right) - \frac{m_e c^2}{4\gamma} \nabla |a|^2.$$

When the pulse velocity is high, and the length is small, then the reverse effect from the plasma on it can be neglected. In this case, it is possible to use the approximation of a given pulse, that is, not to solve a special equation for the envelope, but to use directly the space–time intensity distribution, which, as a rule, varies according to the Gaussian law [29, 73], i.e.

$$a(\rho,\theta) = a_* \exp\left\{-\frac{\rho^2}{\rho_*^2} - \frac{(\theta_{\min} - \theta)^2}{l_*^2}\right\}, \tag{1.29}$$

where a_*, ρ_*, l_* are the amplitude, width and length of the pulse (given parameters), ρ is the distance from the centre of the pulse. If θ_{min} is taken sufficiently large, for example, $\theta_{min} = 4.5\ l_*$, then it follows from formula (1.29) that at the initial time $\theta = 0$, the influence of the pulse is almost absent. Further, as θ increases, the impact intensity first increases (up to $\theta = \theta_{min}$ inclusive), then decreases at the same speed, and, starting from the moment $\theta = 2\ \theta_{min}$, the influence of the pulse again becomes negligible. At the same time, at each instant of time, the intensity distribution over the transverse coordinate is Gaussian in character and rather rapidly decreases in space.

In this consideration, a term of the form $const \cdot \nabla |a|^2$ is added to the right-hand side of the equation for the electron momentum. In particular, such a term modifies the first equation of the P1NE system as follows:

$$\frac{\partial V}{\partial \theta} + E + V \frac{\partial V}{\partial \rho} = \alpha \rho \exp^2 \left\{ -\frac{\rho^2}{\rho_*^2} - \frac{(\theta_{min} - \theta)^2}{l_*^2} \right\},$$

where $\alpha = (a_*/\rho_*)^2$. From this it is easy to conclude that the right-hand side of this form generates particular solutions that have a linear dependence on the spatial variable in the vicinity of the origin and exponentially decrease at some distance from it. Precisely such oscillations are triggered by a laser pulse, moving along a plasma bunch. Therefore, the main object of attention next will be plasma oscillations generated by the initial perturbation of the electric field $\mathbf{E}^0 = const \cdot \nabla |a|^2$, which has the form of a gradient of the intensity function of a laser pulse. The numerical experiments carried out in [45, 123] demonstrated a good qualitative and quantitative correspondence between the processes of breaking of wake waves in a plasma and electronic oscillations generated by a laser pulse with spatial Gaussian intensity.

For definiteness, we note that both the one-dimensional plane electronic oscillations P1RE and P1NE (relativistic (1.24) and non-relativistic (1.25)) systems are supplied with the same initial conditions of the form

$$V^0(\rho) = \beta \rho \exp^2 \left\{ -\frac{\rho^2}{\rho_*^2} \right\}, \quad E^0(\rho) = \alpha \rho \exp^2 \left\{ -\frac{\rho^2}{\rho_*^2} \right\}, \tag{1.30}$$

where α, β are some constants. Moreover, for simplicity, as a rule, we will assume that the oscillations start from a state of rest, that is, $\beta = 0$.

The expressions in the Eulerian variables (1.30) allow us to obtain initial conditions for the integration of independent systems describing the dynamics of particles in the Lagrangian variables (1.27), (1.28). Let us use for this the equation of the particle trajectory with the Lagrangian coordinate ρ^L

$$\rho(\rho^L, \theta) = \rho^L + R(\rho^L, \theta),$$

where R is the displacement relative to the equilibrium position. At the initial time $\theta = 0$, the particle is at the point

$$\rho(\rho^L, 0) = \rho^L + R(\rho^L, 0),$$

and at this point, in accordance with (1.30), an electric field $E(\rho(\rho^L, 0))$ is observed. Note that this field is created as a result of particle displacement. Considering that in the case of plane oscillations the functions of the displacement of the particle and the electric field coincide, we have

$$\rho^L = \rho(\rho^L, 0) - E(\rho(\rho^L, 0)).$$

In other words, we fix the Euler coordinate ρ and observe $E^0(\rho)$, $V^0(\rho)$ in it. At this point, at $\theta = 0$, there is a particle having the Lagrangian coordinate $\rho^L = \rho - E^0(\rho)$. Thus, for this particle from (1.30) the initial conditions follow.

$$R(\rho^L, 0) = E^0(\rho), \quad V(\rho^L, 0) = V^0(\rho).$$

In the case of relativistic equations, the actions leading to initial conditions in Lagrangian variables are similar.

It follows from the above that the physically reasonable formulation of the problem of plasma oscillations, in the first place, is the Cauchy problem. In other words, only the initial conditions defined, generally speaking, in an unbounded spatial domain, should uniquely characterize the solution of the initial equations (1.5)–(1.12).

Summarizing the statement of the simplest problem of electron oscillations, we give a verbal description of the physical process inherent in it. In a certain region occupied by the plasma, in some way (by a laser pulse, an electron bunch, etc.), some electrons are derived from the equilibrium position characterizing a neutral plasma. Displaced relative to the equilibrium position, the electrons form the initial electric field, but they themselves are initially considered to be fixed. The resulting electric field causes the electrons to move to the equilibrium position, but the movement occurs with acceleration. Therefore, having crossed the equilibrium position, similar to the motion of a pendulum, the electrons begin to move in the opposite direction. This is the process of electron plasma oscillations in the field of stationary ions. In principle, it can continue indefinitely, if there are no factors that prevent such a movement. However, after a certain period of time, a breaking may occur, that is, a discontinuous function describing an electric field may form from a smooth function. When simulating a breaking in the Eulerian description of the medium, we will see the electron density singularity; in the Lagrangian description, the intersection of the electron trajectories, which initially occupied a different position in space.

1.4. About boundary conditions

From the results of the previous section, it follows that the need for the boundary conditions can arise only in the case of a study of the problem in the Euler variables; in the case of the Lagrangian variables, they are not required. Therefore, we will discuss the formulation of boundary conditions for differential equations as they are needed.

In addition, in the numerical solution of problems formulated in an unbounded spatial domain, as a rule, it is necessary to use artificial boundary conditions, i.e., some relations for the desired functions defined on the boundary of the final computational domain. The construction of the artificial boundary conditions relies heavily on the properties of the solution to the original problem and is oriented, first of all, to ensure that the given conditions on the border have minimal impact on the solution inside the region, and ideally have no effect whatsoever. The most comprehensive review of ideas and methods for constructing boundary conditions, as well as a detailed bibliography on this subject, is contained in the monograph [60]. However, it was not possible to select a suitable example (analogue) in the case of plasma oscillations in question, so the question of constructing appropriate artificial boundary conditions will be examined in detail a bit later and focused on filling this gap. Here we confine ourselves to the consideration of the simplest variant of the boundary conditions — the conditions for the complete damping of the oscillations, and we will have in mind the statement (1.24), (1.30).

For convenience, we denote the computational domain by the variable ρ as $|\rho| \leq d$ and we will discuss the specification of boundary conditions for $\rho = \pm d$. For some d, we set zero boundary conditions

$$V(\pm d,\theta) = 0, \quad E(\pm d,\theta) = 0, \quad \forall \theta \geqslant 0. \tag{1.31}$$

This approach is incorrect from a formal point of view. The point is that the initial function $E^0(\rho)$ in (1.30) is non-zero on the whole axis OX and setting the conditions of the form (1.31) immediately forms a discontinuous electric field. This means that with $\rho = \pm d$, by virtue of formula (1.23)

$$N(\rho,\theta) = 1 - \frac{\partial E}{\partial \rho},$$

there is a singularity of electron density, i.e. immediate breaking of oscillations.

In numerical analysis, that is, the construction of approximate solutions under the conditions of computational error, this formal obstacle can be overcome as follows. Consider on the interval $[0, h]$, where h is a small parameter (later, the grid mesh), the cubic polynomial $P_3(s)$, which has the following properties:

$$P_3(0) = 1, \quad P_3'(0) = 0, \quad P_3(h) = 0, \quad P_3'(h) = 0.$$

Let's write its explicit expression

$$P_3(s) = \left(\frac{s}{h} - 1\right)^2 \left(\frac{2}{h}s + 1\right).$$

On the interval $[0, h]$, the polynomial is non-negative, its derivative is everywhere non-positive.

Since when finding a smooth classical solution of the problem, we can restrict ourselves to the continuity of its derivatives, it makes sense to consider as an initial condition instead of the function $E^0(\rho)$ a function of the form $\Psi(\rho) = E^0(\rho)\psi(\rho)$, where $\psi(\rho)$ is identically equal to unity with $|\rho| \leq d - h$, on the interval $[d - h, d]$ is defined as $\psi(\rho) = P_3(\rho - d + h)$, and on the interval $[-d, -d + h]$ – symmetrically, i.e. the function $\psi(\rho)$ itself and its derivative at the point $\rho = -d$. Using the monotony of $E^0(\rho)$ and $P(s)$, it is easy to estimate the values

$$\max_{|\rho| \leqslant d} \left| \Psi(\rho) - E^0(\rho) \right| = \left| E^0(d) \right|, \quad \max_{|\rho| \leqslant d} \left| \Psi'(\rho) - \left(E^0 \right)'(\rho) \right| \approx \frac{3}{2h} \left| E^0(d) \right|.$$

Now we introduce on the segment $|\rho| \leq d$ a uniform grid with step h. In all internal nodes of the grid $\rho k = kh$, such that $|\rho k| < d$, the values of $\Psi(\rho k)$ will coincide with the values of $E^0(\rho k)$, and in the boundary nodes, where $\rho = \pm d$, in accordance with (1.31), take zero values. It is easy to see that for any choice of the parameter h, one can indicate such a value d that both the functions $E^0(\rho)$ and $\Psi(\rho)$ themselves, as well as their derivatives, will differ arbitrarily small in the chosen segment $[-d, d]$. In particular, such a choice d is quite sufficient that the estimates given will coincide in order with the computational error, that is, with the rounding errors of the data.

In practice, this restriction is easy to satisfy. For example, setting $d = 4.5\rho_*$, we get $\exp^2\{-d^2 / \rho_*^2\} \approx 2.5768 \cdot 10^{-18}$. This means that in double-precision calculations, the jump in the initial function E^0 is comparable with the machine precision, that is, with the usual error of rounding off the data. In other words, in the numerical simulation of oscillations, the «start» effect of their breaking at such a remote border will not be noticeable at all, which is fully consistent with the concept of an 'artificial boundary'. The above reasoning, by virtue of the form of the initial conditions (1.30), is fully appropriate for the function $V^0(\rho)$. It should be noted that the auxiliary polynomial $P_3(s)$ is required exclusively for theoretical argumentation, since in setting the initial conditions its values on the interval $(0, h)$ are not used in the calculations.

1.5. Bibliography and comments

Plasma is a highly nonlinear medium in which even relatively small initial collective displacements of particles lead to the excitation of oscillations and waves of sufficiently large amplitude. In the absence of dissipation, the evolution of strongly nonlinear oscillations and waves leads to their breaking, i.e., the emergence of a singularity in the function describing the density of the electrons [129].

For a one-dimensional plane non-linear plasma wave, the breaking criterion was found in [15], where the limiting magnitude of the electric field amplitude, to which the wave can exist, was determined, and as we approach the electron density perturbations in the wave become infinitely large.

Fundamentally important is the work [131], in which it was shown that the breaking of oscillations can occur even at amplitudes lower than the limiting value, but after some time after their excitation. It was found that the breaking time increases rapidly with decreasing amplitude of oscillations, since it is inversely proportional to the cube of amplitude [46, 131, 149]. These variations are of primary interest in terms of research.

In [131], the breaking was explained at a qualitative level by the intersection of electron trajectories. In the monograph [59] it was established that the singularity of the density, i.e., the breaking of oscillations, arises as a result of the intersection of the electron trajectories.

In terms of characterizing the qualitative properties of solutions of systems of quasilinear equations, the effect of breaking plasma oscillations that we are interested in is defined as a *gradient catastrophe* [81], that is, the formation of unbounded derivatives (gradients) is meant when the solution itself is bounded. It should be noted that in this case the occurrence of a gradient catastrophe, in essence, means the limit of applicability of the mathematical model under consideration. In other words, if a discontinuous solution of system (1.5)–(1.12) is formed from smooth initial data, for example, continuously differentiable, it means the loss of the physical meaning of the desired functions. Apparently, for this reason, the English-language scientific literature uses the term 'breaking', which embeds the meaning of 'interruption', 'destruction', etc. After observing this effect, further analysis of the fluctuations is not of interest.

It should be noted that the situation described is fundamentally different from modelling in gas dynamics, where discontinuous solutions are physically natural. A consequence of this naturalness is the fact that an exact or approximate solution of the Riemann problem (the Cauchy problem with piecewise discontinuous initial data) is the basis of most modern numerical solution algorithms for the corresponding problem statements [69]. In problems related to plasma oscillations, the formulation of the Riemann problem has no physical meaning, since the initial discontinuous function of the electric field already means an infinite concentration of charge at the points of discontinuity.

When two different particles occupy the same position in space and time, further tracking of their movement requires the involvement of more complex models than classical electrodynamics, since the infinite concentration of the electric charge requires special interpretation. Various plasma models and methods for their investigation are well represented in monographs [53, 115, 139], as well as in the books mentioned above in the chapter.

The author's first acquaintance with plasma models (magnetic hydrodynamics) occurred quite a long time ago [98]. The results of a thorough study of the technology of construction and justification of numerical algorithms for hydrodynamic equations were later described in two monographs: [36] and [99].

The next acquaintance with plasma modelling took place in the framework of joint activities with employees of the Institute of General Physics of the Academy of Sciences of the USSR [18, 19, 28]. The final publications on this topic were articles on physics (on the theory of recombination of X-lasers) [116] and mathematics (on methods for jointly solving the equations of gas dynamics and the kinetics of a multiply charged plasma) [27]. Some of these results were subsequently presented in the monograph [29].

The statements of type 1.2.1 are not fundamentally new. On the contrary, they are often used explicitly or implicitly. For example, one of the equations of Maxwell's electrodynamics is the equation

$$\operatorname{div} \mathbf{B} = 0.$$

However, as a rule, when modelling plasma dynamics, it is clearly not used. The fact is that the 'working' equation is eq. (1.12):

$$\frac{1}{c} \frac{\partial \mathbf{B}}{\partial t} = -\operatorname{curl} \mathbf{E},$$

from which, under the usual assumptions on the smoothness of functions describing the electric and magnetic fields, it follows: if div $\mathbf{B} = 0$ at the initial time, then the magnetic field remains solenoidal at an arbitrary time. In other words, when considering sufficiently smooth solutions of the equations of magnetic hydrodynamics, one should keep in mind the structure of the initial data for them; often this simplifies the model under consideration.

For example, it follows from Statement 1.2.1 that under the initial condition (1.14), for a sufficiently smooth solution of system (1.5)–(1.12), the following equality is valid

$$\mathbf{B}(\mathbf{x},t) + \frac{c}{e} \operatorname{curl} \mathbf{p}_e(\mathbf{x},t) = 0. \tag{1.32}$$

This relation is well known for stationary nonrelativistic equations, (see, for example, [37], [78]) and is associated, first of all, with irrotational plasma motion. For nonstationary relativistic equations, relation (1.32), apparently, first appeared in [3] and was called the 'law of conservation of a generalized vorticity'. In this case, the above statement pursues an exclusively practical goal — to somewhat simplify the original equations for the further asymptotic, analytical, and numerical study of plasma oscillations generated by the initial conditions in the form of a gradient of a certain function.

Another way to prove Statement 1.2.1, similar to the conclusion that the magnetic field is 'frozen' into a plasma, is given in Ref. [104].

In the above proof of Statement 1.2.1, we used the Gronwall inequality in differential form. We give the formulation of this inequality, following [64]:

Let y be a function that is absolutely continuous on the interval [0, T], *p, q be functions summable on* [0, T], *and let inequality be true almost everywhere on* [0, T]

$$y'(t) \leqslant p(t)y(t) + q(t).$$

Then

$$y \leqslant e^{p_1(t)} \left(y(0) + \int_0^t q(\tau)e^{-p_1(\tau)}d\tau \right) \quad \forall t \in [0,T],$$

where $p_1(t) = \int^t p(\tau)d\tau.$

It should be noted that a linear change in the desired functions in a neighbourhood of the origin of coordinates leads, for problems describing plasma oscillations, to interesting objects that are conveniently called 'axial solutions'. Such solutions (further special attention will be given to them) have an independent physical meaning and are regularly reviewed in the literature (see, for example, [100, 107, 120, 143]).

The described approach to the construction of artificial boundary conditions – the 'cutting off' of an infinite region using homogeneous boundary conditions of the first kind – is practically quite convenient and therefore often used. However, the main disadvantage of this approach is an excessive increase in the computational domain. The desired breaking effect is usually observed in the vicinity of the beginning of the coordinate ρ at a distance of less than 0.1 ρ_*, therefore more than 90% of the calculations are a kind of 'fee' for not knowing the appropriate boundary conditions. A separate presentation will be devoted to other approaches to the construction of artificial boundary conditions in the simulation of breaking. The most comprehensive review of ideas and methods for constructing boundary conditions of this kind, as well as a detailed bibliography on this subject, is contained in the monograph [60].

When formulating problems, the principle of symmetry of physical processes is often useful, which plays an important role in formulating the initial and boundary conditions for the initial system (1.5)–(1.12). If, for example, the initial conditions for the P1NE system are odd functions and a solution is of interest, which has the oddness property relative to the origin of coordinates over the entire interval of their existence, then for computational needs, even in the simplest case of using the Cartesian coordinate system, this reduces the computational area by half. Naturally, what has been said is fully applicable to the parity property of the solution. If a cylindrical (or spherical) coordinate system is used, the oscillation symmetry gives grounds for the formulation of boundary conditions on the axis of symmetry of the computational domain.

2

Plane one-dimensional non-relativistic electron oscillations

The chapter deals with plane one-dimensional non-relativistic electron oscillations described by the P1NE system of equations. The main goal is to clarify the conditions under which such oscillations can exist indefinitely. Of particular interest are the 'axial solutions' of the P1NE system, as well as solutions of a special type built on their basis.

2.1. Problem statement in Eulerian and Lagrangian variables

We will study one of the simplest formulations of the initial–boundary value problem for the P1NE equations, which describes free plane one-dimensional non-relativistic oscillatory electron motions. We recall her.

It is required in the semi-infinite strip $\{(\rho, \theta): -d < \rho < d, \theta > 0\}$ to find a solution to the equations

$$\frac{\partial V}{\partial \theta} + E + V\frac{\partial V}{\partial \rho} = 0, \quad \frac{\partial E}{\partial \theta} - V + V\frac{\partial E}{\partial \rho} = 0, \tag{2.1}$$

satisfying the initial

$$V(\rho,0) = V_0(\rho), \quad E(\rho,0) = E_0(\rho), \quad \rho \in [-d,d], \tag{2.2}$$

and boundary conditions

$$V(-d,\theta) = V(d,\theta) = E(-d,\theta) = E(d,\theta) = 0 \quad \forall \theta \geq 0. \qquad (2.3)$$

In this formulation, we consider oscillations of the plasma stab, since the boundary conditions (2.3) prevent any perturbations from moving beyond the boundaries of the segment $[-d, d]$.

Where specifics are required, we will keep in mind the initial functions of the form

$$V_0(\rho) = \beta \rho \exp^2\left\{-\frac{\rho^2}{\rho_*^2}\right\}, \quad E_0(\rho) = \alpha \rho \exp^2\left\{-\frac{\rho^2}{\rho_*^2}\right\}$$

with some constants α, β, as well as considerations related to the 'cut-off' of the region under consideration at $d = 4.5\rho_*$. In the general case, we will discuss sufficiently smooth (as a minimum, continuously differentiable) solutions and the corresponding initial conditions that are smoothly adjacent to the boundary ones.

It should be noted that the index '0', which marks the initial conditions, in this chapter, for convenience of notation, is located below the functions.

The quasilinear system of equations (2.1) is the main one in the chapter, therefore, in addition to writing in the Eulerian variables, its form in the Lagrangian variables will be useful:

$$\frac{dV(\rho^L,\theta)}{d\theta} = -E(\rho^L,\theta), \quad \frac{dE(\rho^L,\theta)}{d\theta} = V(\rho^L,\theta),$$

where $d/d\theta = \partial/\partial\theta + V\,\partial/\partial\rho$ is the total time derivative.

Recall that the quantity R, which determines the displacement of a particle with the Lagrangian variable ρ^L,

$$\rho(\rho^L,\theta) = \rho^L + R(\rho^L,\theta), \qquad (2.4)$$

satisfies the equation

$$\frac{dR(\rho^L,\theta)}{d\theta} = V(\rho^L,\theta).$$

From this it follows that the values of $R(\rho^L, \theta)$ and $E(\rho^L, \theta)$ coincide up to a constant, which is calculated from the condition that the electric field is zero in the absence of displacements. In other words, in the case of plane one-dimensional oscillations, the relation $R(\rho^L, \theta) \equiv E(\rho, \theta)$ is valid, and the basic system of equations (2.1) in the Lagrangian variables takes the form

$$\frac{dV}{d\theta} = -R, \quad \frac{dR}{d\theta} = V. \tag{2.5}$$

We note that relation (2.4) is very useful for determining the Lagrangian coordinate of a particle ρ^L from the initial distribution $E_0(\rho)$. Moreover, the formula is as follows:

$$\rho^L = \rho(\rho^L, 0) - E_0(\rho(\rho^L, 0)). \tag{2.6}$$

To summarize the above, the trajectories of all particles, each of which is identified by the Lagrangian coordinate ρ^L, can be determined by independent integration of the system of ordinary differential equations (2.5). For this, two initial conditions are required: $R(\rho^L, 0)$ and $V(\rho^L, 0)$. To determine $R(\rho^L, 0)$, first set the position of the particle at the initial time ρ, then the displacement at this point is given by the electric field $R(\rho^L, 0) = E_0(\rho)$. In turn, the Lagrangian coordinate of a particle is determined by formula (2.6). In a similar way, $V(\rho^L, 0)$ is defined by $V_0(\rho)$. Knowledge of the Lagrangian coordinate ρ^L and the displacement function $R(\rho^L, \theta)$ uniquely characterizes the particle trajectory by formula (2.4).

2.2. Axial solutions

In [100], for nonlinear problems in cylindrical geometry, describing laser–plasma interactions and possessing axial symmetry, the concept of an axial solution was introduced as a solution having a locally linear dependence on the spatial coordinate. In this (plane) case, axial symmetry, of course, is not assumed, but for convenience, we will use this name for real solutions of system (2.1) of the form

$$V(\rho, \theta) = W(\theta)\rho, \quad E(\rho, \theta) = D(\theta)\rho$$

and study some of their properties.

It is easy to make sure that the time-dependent factors in this case satisfy the system of ordinary differential equations.

$$W' + D + W^2 = 0, \quad D' - W + WD = 0. \tag{2.7}$$

We supplement the equations obtained with arbitrary real initial conditions

$$W(0) = \beta, \quad D(0) = \alpha \tag{2.8}$$

and find out the conditions for the existence and uniqueness of the solution of the Cauchy problem (2.7), (2.8).

Note that the reduced Cauchy problem is not trivial, since it admits both regular 2π-periodic solutions (for example, for small α and β) and solutions that have singularities on a finite time interval (the so-called blow-up solutions). Moreover, the available results even for polynomial right-hand sides (see [137]) in the general case do not allow us to establish the exact boundary between the sets of initial data generating solutions of various types. Therefore, it seems useful to have a slightly different look at the problem under consideration. The following takes place:

Lemma 2.2.1. *The Cauchy problem* (2.7), (2.8) *is equivalent to the following differential algebraic problem:*

$$W' = (1 - 2\alpha - \beta^2)x^2 + (\alpha - 1)x, \tag{2.9}$$

$$W^2 + 1 + (1 - 2\alpha - \beta^2)x^2 + 2(\alpha - 1)x = 0, \tag{2.10}$$

$$W(0) = \beta, \quad x(0) = 1. \tag{2.11}$$

Proof. After excluding the function D from system (2.7), we obtain the Cauchy problem for a second-order equation:

$$W'' + 3W'W + W + W^3 = 0, \quad W(0) = \beta, \quad W'(0) = -(\alpha + \beta^2). \tag{2.12}$$

We lower the order of the equation by replacing $p(W) = W'_\theta$:

$$p'_W p + 3pW + W + W^3 = 0. \tag{2.13}$$

Hereinafter, the subscript of the derivative clearly indicates the independent variable by which differentiation is carried out. Note that equation (2.13) corresponds (see problem (2.12)) to the initial condition

$$p(\beta) = -(\alpha + \beta^2). \tag{2.14}$$

Further, the transformation of the dependent variable $p(W) = u^{-1}(W) \neq 0$ leads to the equation

$$u_W' - 3u^2 W - u^3(W + W^3) = 0,$$

in which it is convenient to make the substitution $u(W) = \eta(\xi)$, where $\xi = 3W^2/2 + C_\xi$. As a result, we will have

$$\eta_\xi' = g(\xi)\eta^3 + \eta^2, \quad g(\xi) = \frac{2}{9}\xi + \frac{1}{3}\left(1 - \frac{2}{3}C_\xi\right).$$

To obtain an analytical solution of this equation, we introduce the parameterization of the independent variable $\xi = \xi(t)$ so that

$$\xi_t' = -\frac{1}{t\eta(\xi)}, \quad t \neq 0,$$

and as its corollary we come to the equation

$$t^2\xi_t'' + \frac{2}{9}\xi + \frac{1}{3}\left(1 - \frac{2}{3}C_\xi\right) = 0.$$

Its general solution is

$$\xi(t) = C_1 t^{2/3} + C_2 t^{1/3} + C_\xi - \frac{3}{2},$$

where does the formula come from

$$\eta(\xi) = -\left(\frac{2}{3}C_1 t^{2/3} + \frac{1}{3}C_2 t^{1/3}\right)^{-1},$$

since $\eta(\xi) = -[t\,\xi_t'(t)]^{-1}$. Considering that $p(W) = u^{-1}(W)$, where $u(W) = \eta(\xi)$, returning to the original variables gives

$$p(W) = -\left(\frac{2}{3}C_1 t^{2/3} + \frac{1}{3}C_2 t^{1/3}\right), \quad W^2 + 1 = \frac{2}{3}\left(C_1 t^{2/3} + C_2 t^{1/3}\right).$$

To derive the second formula, the relation $\xi = 3W^2/2 + C_\xi$ was used here, i.e.

$$W^2 = \frac{2}{3}(\xi - C_\xi) = \frac{2}{3}\left(C_1 t^{2/3} + C_2 t^{1/3}\right) - 1.$$

And finally, the definition of the constant C_1, C_2 from condition (2.14), that is, from the agreement of the values of the parameters $\theta = 0$ and $t = 1$, and the formal replacement of $t^{1/3} = x$ lead to the differential–algebraic problem (2.9) – (2.11). The lemma is proved.

Note that the procedure performed is an uncomplicated sequence of well-known techniques (see, for example, [62]). Its practical use in this case consists in obtaining an equivalent problem not only in a closed but also in a form more convenient for research. It should, of course, be clarified that in this case the equivalence of the considered statements is understood as follows: the function $W(\theta)$ in both statements is the same, and the remaining functions $D(\theta)$ and $x(\theta)$ are determined in each statement by $W(\theta)$ uniquely. With the help of the proven statement it is established:

Theorem 2.2.1. *A necessary and sufficient condition for the existence and uniqueness of a smooth periodic solution of the Cauchy problem* (2.7), (2.8) *is the fulfillment of the inequality*

$$1 - 2\alpha - \beta^2 > 0. \tag{2.15}$$

Proof. *Sufficiency.* Let the expression on the left side of inequality (2.15) be strictly positive, denote it by c^2. In this case, the manifold (2.10) is an ellipse

$$\frac{W^2}{a^2} + \frac{y^2}{b^2} = 1,$$

where

$$a^2 = \frac{\alpha^2 + \beta^2}{c^2}, \quad b^2 = \frac{a^2}{c^2}, \quad y = x + \frac{\alpha - 1}{c^2},$$

and equation (2.9) takes the form

$$W' = c^2 y^2 - (\alpha - 1)y.$$

Now, from the smoothness and compactness of the manifold (2.10), as well as from the smoothness of the right-hand side of (2.7) in the normal form of the equation, there follows the existence and uniqueness of a smooth solution that can be infinitely extended in both directions with respect to $\theta = 0$ (see, for example, [13]) and the Poincaré – Bendixson theorem [66] guarantees its periodicity. Sufficiency is proven.

Necessity. Suppose inequality (2.15) is not satisfied, then two cases are possible. Consider them sequentially.

First, let $1 - 2\alpha - \beta^2 = 0$, then the manifold (2.10) is a parabola $W^2 = 2y$, where $y = (1 - \alpha) x - 1/2$, and the corresponding equation (2.9) has the form

$$W' = -\left(y + \frac{1}{2}\right) = -\frac{1}{2}(W^2 + 1).$$

The solution of this equation with the initial condition W (0) = β is easy to write explicitly:

$$W_0(\theta) = \frac{\beta - \mathrm{tg}(\theta/2)}{1 + \beta\,\mathrm{tg}(\theta/2)},$$

whence it follows that it monotonously decreases and becomes unbounded for some $0 < \theta_* < 2\pi$ such that either the denominator is zero for $\beta \neq 0$ or $\theta_* = \pi$ for $\beta = 0$. In conclusion of the consideration of this case, we note that the equalities $\alpha = (1 - \beta^2)/2$ and $\alpha = 1$ are incompatible.

Now suppose that the expression on the left side of inequality (2.15) is strictly negative, we introduce the notation $(-c^2)$ for it. In this case, the manifold (2.10) is a hyperbola

$$\frac{y^2}{b^2} - \frac{W^2}{a^2} = 1,$$

where $y = x - (\alpha - 1)/c^2$, and the values a^2 and b^2 have the same form as above. In this case, equation (2.9) takes the form

$$W' = -c^2 y^2 - (\alpha - 1)y.$$

We show that with an arbitrary initial condition $W(0) = \beta$, the solution of this equation decreases faster than $W_0(\theta)$. By the comparison theorem, it suffices to establish the validity of the inequality

$$-c^2 y^2 - (\alpha - 1)y < -\frac{1}{2}(W^2 + 1) \qquad (2.16)$$

on the hyperbola under consideration for any values of W. Inequality (2.16) with $y = \pm b(1 + W^2/a^2)^{1/2}$ is equivalent to the inequality

$$\frac{W^2}{2} + \frac{(\alpha - 1/2)^2 + \beta^2/2 + 1/4}{c^2} + (\alpha - 1)y > 0.$$

Therefore, if the term $(\alpha - 1)y$ is nonnegative, then inequality (2.16) is obviously true. Otherwise, by transferring the indicated addend to the right side and squaring both sides, after elementary transformations we get the inequality

$$\left(\frac{W^2}{2} + \frac{1}{2} \right)^2 > 0,$$

which also guarantees the validity of the majorant estimate (2.16) for any values of W. Finally we conclude that if the condition (2.15) is not fulfilled, the solution is unbounded on a finite time interval, that is, the blow-up property. The theorem is proved.

Let us discuss the result from two points of view: the physical interpretation of functions and finding the approximate (numerical) solution of the problem (2.7), (2.8).

For the Cauchy problem under study, we consider the simplest unbounded solution $W_0(\theta)$ that satisfies the equation $W' = -(W^2 + 1)/2$. This implies the formula $D_0(\theta) = (1 - W_0^2(\theta))/2$, and, accordingly, the expression for the electron density $N_0 = 1 - D_0 = (W_0^2 + 1)/2$. Thus, from the physical point of view, $W \to -\infty$, $D \to -\infty$ means $N \to +\infty$, i.e., an unlimited increase in the electron density when condition $1 - 2\alpha - \beta^2 \leq 0$ is fulfilled. Note that in this situation the critical is the value $\alpha_* = 1/2$, i.e. $N_* = 1/2$. In other words, a decrease in the initial electron density by a factor of two (or more!) compared to the unperturbed value ($N \equiv 1$) already guarantees a blow-up solution for any initial velocity distribution, which is characterized by the parameter β.

From the point of view of numerical methods, it should be noted that the differential-algebraic formulation (2.9)–(2.11) has no visible advantages compared with the fully differential formulation (2.7), (2.8). On the contrary, the Cauchy problem (2.7), (2.8), which has a smooth periodic solution, can be successfully numerically integrated by a variety of methods (see, for example, [16, 22]), but the convergence of an approximate solution to the exact one can be accurately justified only under the condition of the stability of the solution with respect to the initial data, which is closely related to the ellipticity of the manifold (2.10).

We give a useful presentation of the solution.

Statement 2.2.1. *The 2π-periodic solution of the Cauchy problem* (2.7), (2.8) *under condition* (2.15) *is given by the formulas*

$$W(\theta) = a\sin\varphi(\theta), \quad D(\theta) = -(a^2 + af\cos\varphi(\theta)), \qquad (2.17)$$

where

$$f = \sqrt{1+a^2}, \quad \varphi(\theta) = 2\operatorname{arc\,ctg}\left[\frac{1}{a+f}\,\operatorname{tg}\left(C_0 - \frac{\theta}{2}\right)\right].$$

In this case, the constant C_0 and the initial angle $\varphi_0 = \varphi(0)$ are determined by the formulas

$$C_0 = \operatorname{arc\,tg}\left[(a+f)\operatorname{ctg}\left(\varphi_0 / 2\right)\right], \quad \sin\varphi_0 = \frac{\beta}{a}, \quad \cos\varphi_0 = -\frac{\alpha+\beta^2}{ac},$$

and the remaining values have the same meaning:

$$c = \sqrt{1 - 2\alpha - \beta^2}, \quad a = \frac{\sqrt{\alpha^2 + \beta^2}}{c}, \quad b = \frac{a}{c}.$$

Proof. We use the ellipticity of the manifold (2.10) under condition (2.15): we set

$$W = a\sin\varphi, \quad y = b\cos\varphi.$$

Then from (2.9) we obtain the equation for the function $\varphi(\theta)$:

$$\varphi'_\theta = a\cos\varphi + f, \quad f = b(1-\alpha)/a. \qquad (2.18)$$

Since $f > a$, the solution (2.18) has the form (see [62])

$$\operatorname{arc tg}\left(\sqrt{\frac{f+a}{f-a}}\operatorname{ctg}\frac{\varphi}{2}\right)+\frac{\theta}{2}\sqrt{f^2-a^2}=C_0.$$

Simple calculations lead to relations

$$f^2-a^2=1,\quad \beta=a\sin\varphi_0,\quad -\frac{\alpha+\beta^2}{c^2}=b\cos\varphi_0,$$

where, after determining the value of φ_0, follows the expression for C_0. Recall that from (2.7) it follows that $D = -(W' + W^2)$; this, in turn, leads to an explicit formula for $D(\theta)$. The statement is proven.

Using the Lagrangian variables, we can derive other formulas for axial solutions:

$$W(\theta)=\frac{s\cos(\theta+\theta_0)}{1+s\sin(\theta+\theta_0)},\quad D(\theta)=\frac{s\sin(\theta+\theta_0)}{1+s\sin(\theta+\theta_0)},\qquad (2.19)$$

where

$$s=\frac{\sqrt{\alpha^2+\beta^2}}{1-\alpha},\quad \cos\theta_0=\frac{\beta}{\sqrt{\alpha^2+\beta^2}},\quad \sin\theta_0=\frac{\alpha}{\sqrt{\alpha^2+\beta^2}}.$$

Their identity with formulas (2.17) is not obvious, however, it is not difficult to verify.

Perform the derivation of new formulas step by step.

1. For a particle that is characterized by a Lagrangian variable ρ^L, the following equations are valid

$$\frac{dV(\rho^L,\theta)}{d\theta}=-R(\rho^L,\theta),\quad \frac{dR(\rho^L,\theta)}{d\theta}=V(\rho^L,\theta).$$

This implies:

$$R(\rho^L,\theta)=A\sin(\theta+\theta_0),V(\rho^L,\theta)=A\cos(\theta+\theta_0),$$
$$\text{where}\quad A=A(\rho^L),\theta_0=\theta_0(\rho^L).\qquad (2.20)$$

Here the values $|A|$ and θ_0 have the meaning of the amplitude and initial phase of the oscillations.

2. Recall that for the equations in the Eulerian variables

$$\frac{\partial V}{\partial \theta} + E + V \frac{\partial V}{\partial \rho} = 0, \quad \frac{\partial E}{\partial \theta} - V + V \frac{\partial E}{\partial \rho} = 0$$

the axial solutions are sought

$$V(\rho,\theta) = W(\theta)\rho, \quad E(\rho,\theta) = D(\theta)\rho,$$

where the initial conditions are given by real values:

$$W(0) = \beta, \quad D(0) = \alpha, \quad \alpha^2 + \beta^2 \neq 0.$$

3. We define the connection between the Lagrangian variable ρ^L and its value of the Eulerian variable ρ at the initial time $\theta = 0$. We write the formula for the particle trajectory

$$\rho(\rho^L,\theta) = \rho^L + R(\rho^L,\theta)$$

and use the equality $R(\rho^L, \theta) = E(\rho, \theta)$. At the initial moment of time we get

$$\rho \equiv \rho(\rho^L,0) = \rho^L + E_0(\rho(\rho^L,0)), \quad E_0(\rho) = \alpha\rho,$$

what gives

$$\rho^L = \rho - E_0(\rho) = \rho(1-\alpha), \quad \text{or} \quad \rho = \frac{\rho^L}{1-\alpha}.$$

Note that the preservation of the order of the particles during the formation of a given initial electric field is guaranteed by the condition $1 - \alpha > 0$.

4. Determine the amplitude of oscillations. Since, by virtue of (2.20), the value

$$A^2(\rho^L) = R^2(\rho^L,\theta) + V^2(\rho^L,\theta)$$

depends only on ρ^L, then it can be determined at $\theta = 0$:

$$A^2(\rho^L) = E_0^2(\rho(\rho^L,0)) + V_0^2(\rho(\rho^L,0)) =$$

$$= E_0^2\left(\frac{\rho^L}{1-\alpha}\right) + V_0^2\left(\frac{\rho^L}{1-\alpha}\right) = \left(\rho^L\right)^2 \frac{\alpha^2 + \beta^2}{(1-\alpha)^2}.$$

From here we have

$$A(\rho^L) = \rho^L \frac{\sqrt{\alpha^2 + \beta^2}}{1-\alpha} \equiv \rho^L s.$$

5. Determine the initial phase of the oscillations $\theta_0 = \theta_0(\rho^L)$. At the initial moment of time we have

$$R(\rho^L,0) = A(\rho^L)\sin\theta_0,$$

on the other hand

$$R(\rho^L,0) = E(\rho,0) = E_0(\rho) = \alpha\rho = \alpha \frac{\rho^L}{1-\alpha}.$$

This implies

$$\sin\theta_0 = \alpha \frac{\rho^L}{1-\alpha} A^{-1}(\rho^L) = \frac{\alpha}{\sqrt{\alpha^2 + \beta^2}}.$$

Similarly, from the formula $V(\rho^L, 0) = A(\rho^L)\cos\theta_0$, we obtain

$$\cos\theta_0 = \frac{\beta}{\sqrt{\alpha^2 + \beta^2}}.$$

6. Final formulas for $D(\theta)$ and $W(\theta)$. From relations

$$R(\rho^L,\theta) = E(\rho,\theta) \equiv D(\theta)\rho(\rho^L,\theta)$$

follows

$$D(\theta) = \frac{R(\rho^L, \theta)}{\rho(\rho^L, \theta)} = \frac{\rho^L s \sin(\theta + \theta_0)}{\rho(\rho^L, \theta)}.$$

Let's present in a convenient form a denominator:

$$\rho(\rho^L, \theta) = \rho^L + R(\rho^L, \theta) = \rho^L (1 + s \sin(\theta + \theta_0)),$$

and write the final formula:

$$D(\theta) = \frac{s \sin(\theta + \theta_0)}{1 + s \sin(\theta + \theta_0)}.$$

The formula for $W(\theta)$ differs from $D(\theta)$ by replacing $\sin(\theta + \theta_0)$ with $\cos(\theta + \theta_0)$ in the numerator in accordance with (2.20).

The correctness of the formulas $D(\theta)$ and $W(\theta)$ is also related to condition (2.15), which is easy to verify. Indeed, consider the denominator of both formulas: $1 + s \sin(\theta + \theta_0)$. From its explicit form, it follows that the regularity condition for oscillations coincides with the inequality $|s| < 1$, i.e.

$$-1 < \frac{\sqrt{\alpha^2 + \beta^2}}{1 - \alpha} < 1.$$

In addition, it follows from the conservation of the order of the particles $\rho^L \rho > 0$ and the formula $\rho^L = \rho(1 - \alpha)$ that $1 - \alpha > 0$. Hence we have

$$\sqrt{\alpha^2 + \beta^2} < 1 - \alpha, \quad \text{or} \quad 1 - 2\alpha - \beta^2 > 0.$$

In conclusion of the consideration of formulas (2.19), we note that they satisfy relations (2.7), (2.8), therefore, by virtue of the proved theorem of existence and uniqueness, as well as (2.17), represent the desired axial solution.

We note that condition (2.15) is conserved in time, that is, from formulas (2.19) for any θ, follows the inequality $1 - 2D(\theta) - W^2(\theta) > 0$. Indeed, for convenience, we denote $s \sin(\theta + \theta_0)$ by p, then

$$1-2D(\theta)-W^2(\theta)=1-\frac{2p}{1+p}-\frac{s^2(1-\sin^2(\theta+\theta_0))}{(1+p)^2}=$$

$$=1-\frac{2p}{1+p}-\frac{s^2-p^2}{(1+p)^2}=\frac{1-s^2}{(1+p)^2}=\frac{1-2\alpha-\beta^2}{(1-\alpha)^2}>0.$$

Similarly, the preservation in time of condition (2.15) is easy to verify using formulas (2.17).

2.3. 'Triangular' solutions

The results of the previous section allow to construct in analytical form piecewise linear ('triangular') solutions of the initial–boundary value problem for the main system of equations (2.1)

$$\frac{\partial V}{\partial \theta}+E+V\frac{\partial V}{\partial \rho}=0,\quad \frac{\partial E}{\partial \theta}-V+V\frac{\partial E}{\partial \rho}=0.$$

Let the initial functions (2.2) have the form of triangles:

$$V_0(\rho)=\begin{cases}\beta_1(\rho+d), & -d\le\rho\le\rho_V^0,\\ \beta_2(\rho-d), & \rho_V^0\le\rho\le d,\end{cases}$$

$$E_0(\rho)=\begin{cases}\alpha_1(\rho+d), & -d\le\rho\le\rho_E^0,\\ \alpha_2(\rho-d), & \rho_E^0\le\rho\le d,\end{cases} \tag{2.21}$$

with two fixed vertices ($\pm d$, 0) and ρ-projections of the third vertex ρ_V^0,ρ_E^0, respectively. In the general case, the quantities ρ_V^0,ρ_E^0 belong to the interval ($-d$, d), therefore the parameters of the triangles, by virtue of continuity at these points, are connected by the relations

$$\beta_1(\rho_V^0+d)=\beta_2(\rho_V^0-d),\quad \alpha_1(\rho_E^0+d)=\alpha_2(\rho_E^0-d).$$

It is also possible that one of the initial functions of E_0 or V_0 is identically zero, then we assume that the values of ρ_V^0 and ρ_E^0 coincide.

2.3.1. Simple solutions

We will 'paste together' the axial solutions in the simplest case ρ_V^0 and ρ_E^0 are equal to each other. Denote the unique breakpoint of the derivatives by $\rho_I^0 = \rho_V^0 = \rho_E^0$ and define the 'triangular' solution as follows:

$$V(\rho,\theta) = \begin{cases} W_1(\theta)(\rho+d), & -d \le \rho \le \rho_I(\theta), \\ W_2(\theta)(\rho-d), & \rho_I(\theta) \le \rho \le d, \end{cases}$$

$$E(\rho,\theta) = \begin{cases} D_1(\theta)(\rho+d), & -d \le \rho \le \rho_I(\theta), \\ D_2(\theta)(\rho-d), & \rho_I(\theta) \le \rho \le d, \end{cases} \tag{2.22}$$

where $W_i(\theta)$, $D_i(\theta)$, $i = 1, 2$, are axial solutions of the basic system (2.1) with the corresponding initial data β_i, α_i, $i = 1, 2$, satisfying condition (2.15), and the function $\rho_I(\theta)$ is given by the continuity conditions at the breakpoint of the derivatives of the functions $V(\rho, \theta)$ and $E(\rho, \theta)$.

To write $\rho_I(\theta)$ explicitly, we use the main system in the Lagrangian coordinates. First, in accordance with (2.6), we define the Lagrangian coordinate of the particle, which at $\theta = 0$ is at the point ρ_I^0:

$$\rho_I^L = \rho_I^0 - E_0(\rho_I^0).$$

Further, from equations (2.5) we obtain

$$R(\rho_I^L,\theta) = A(\rho_I^L)\sin(\theta + \theta_I^L), \quad V(\rho_I^L,\theta) = A(\rho_I^L)\cos(\theta + \theta_I^L),$$

where

$$A(\rho_I^L) = \sqrt{V_0^2(\rho_I^0) + E_0^2(\rho_I^0)},$$
$$\cos\theta_I^L = V_0(\rho_I^0) / A(\rho_I^L), \quad \sin\theta_I^L = E_0(\rho_I^0) / A(\rho_I^L).$$

This gives the desired analytical dependence

$$\rho_I(\theta) = \rho_I^L + R(\rho_I^L,\theta). \tag{2.23}$$

It is easy to verify that solution (2.22) satisfies the basic system (2.1) on both sides of the moving break (2.23), and is continuous on the break itself. This means that the above formulas describe a local in space (on the segment $[-d, d]$) 2π-periodic in time solution, which at any moment has the form of two triangles, each of which degenerates with period π into a segment.

2.3.2. Composite solutions

We now consider a more general case of a possible distinction of the quantities ρ_V^0 and ρ_E^0. To this end, we divide the interval $[-d, d]$ into three non-empty non-intersecting parts (otherwise, the 'triangular' solutions will be simple) and for each of the parts we define from the initial functions (2.2) its own parameters of axial solutions (α_i, β_i), $i = 1, 2, 3$. For example, in the case of $\rho_V^0 < \rho_E^0$ we will have: (α_1, β_1), (α_1, β_2), (α_2, β_2). In such a situation (in the presence of three axial solutions $W_i(\theta)$, $D_i(\theta)$) for 'pasting together' solutions from the continuity conditions, it is necessary to have two equations for the moving break points of the derivatives:

$$\rho_E(\theta) = \rho_E^L + R(\rho_E^L, \theta), \quad \rho_V(\theta) = \rho_V^L + R(\rho_V^L, \theta),$$

each of which is defined similarly to (2.23).

However, we note that in the general case 2π-periodic in time composite solutions may not exist for two reasons: firstly, one of the sets of parameters (α_i, β_i) may not satisfy condition (2.15), and secondly, change the initial order of the Lagrangian particles, i.e., some of their trajectories can intersect. Recall that the intersection of particle trajectories in the Lagrangian description leads at least to the violation of the adequacy of the hydrodynamic model [131], and in the Eulerian variables this situation means that the electron density becomes infinite [59]. We show that the fulfillment of the necessary and sufficient condition for the existence of axial solutions (2.15) allows us to avoid both problems simultaneously.

So, let condition (2.15) be satisfied for any of the three sets (α_i, β_i), then we will discuss the possibility of intersection of the particle trajectories. We write the system (2.5) in the form

$$\frac{d^2 R(\rho^L, \theta)}{d\theta^2} + R(\rho^L, \theta) = 0 \tag{2.24}$$

and we will give its general solution:

$$R(\rho^L, \theta) = R_1(\rho^L)\sin\theta + R_2(\rho^L)\cos\theta. \qquad (2.25)$$

Lagrangian particle coordinate ρ^L and the Cartesian coordinate ρ are initially related by

$$\rho^L = \begin{cases} \rho(1-\alpha_1) - \alpha_1 d, & -d \le \rho \le \rho_E^0, \\ \rho(1-u_2) + u_2 d, & \rho_E^0 \le \rho \le d. \end{cases} \qquad (2.26)$$

Since we have already fixed that $\alpha_i < 1/2$ (due to the validity of condition (2.15)), this function is uniquely reversible:

$$\rho = \begin{cases} (\rho^L + \alpha_1 d)/(1-\alpha_1), & -d \le \rho^L \le \rho_E^L, \\ (\rho^L - \alpha_2 d)/(1-\alpha_2), & \rho_E^L \le \rho \le d. \end{cases} \qquad (2.27)$$

We now consider relation (2.4) for two infinitely close particles with the coordinates ρ^L and $\rho^L + \Delta\rho^L$, respectively. Suppose that their Euler coordinates become equal at some time θ. Then, under the condition of differentiability of $R(\rho^L, \theta)$, we obtain the equality

$$\frac{\partial R(\rho^L, \theta)}{\partial \rho^L} = -1.$$

On the contrary, so that the particle trajectories do not intersect, it suffices to satisfy

$$\left| \frac{\partial R(\rho^L, \theta)}{\partial \rho^L} \right| < 1,$$

which, in turn, follows from the restriction on the total energy of the particle:

$$T(\rho^L, \theta) \equiv \left(\frac{\partial R(\rho^L, \theta)}{\partial \rho^L} \right)^2 + \left(\frac{\partial V(\rho^L, \theta)}{\partial \rho^L} \right)^2 < 1. \qquad (2.28)$$

It is easy to see that the value $T(\rho^L, \theta)$, in fact, does not depend on time, since the function $\partial R(\rho^L, \theta)/\partial \rho^L$ simultaneously with $R(\rho^L, \theta)$ satisfies equation (2.24). A similar statement holds for the functions

$\partial V(\rho^L, \theta)/\partial \rho^L$ and $V(\rho^L, \theta)$, which follows from equations (2.5). Therefore, to verify condition (2.28), it suffices to analyze only the value of $T(\rho^L, 0)$. Using the analytical formulas for the initial functions (2.21), as well as the dependence $\rho = \rho(\rho^L)$ (2.27), we calculate

$$T(\rho^L,0)=\left(\frac{\partial E_0}{\partial \rho}\frac{\partial \rho}{\partial \rho^L}\right)^2+\left(\frac{\partial V_0}{\partial \rho}\frac{\partial \rho}{\partial \rho^L}\right)^2=\frac{\alpha_i^2}{(1-\alpha_i)^2}+\frac{\beta_i^2}{(1-\alpha_i)^2},$$

where the set of the parameters (α_i, β_i) is uniquely determined by the Lagrangian coordinate ρ^L. As a result, it is established that the inequality $T(\rho^L, 0) < 1$ is equivalent to the condition $1-2\alpha_i-\beta_i^2>0$ and this completes the argument. Thus, we have obtained that if condition (2.15) is satisfied at each segment of the simultaneous linearity of the initial functions (2.21), then this guarantees the existence of a 2π-periodic in time composite triangular solution.

Based on the obtained representation for composite 'triangular' solutions, it can be concluded that the maximum possible amplitude of oscillations A_{max} is strictly less than the half width d of the plasma layer under consideration, that is, $A_{max} = d - \varepsilon$ with arbitrarily small $\varepsilon > 0$, and A_{max} is achieved on a simple 'triangular' solution defined by the following initial parameters in (2.21):

$$\rho_I^L=0, \quad \alpha_1=\alpha_2=0, \quad \beta_1=-\beta_2=\pm(1-\varepsilon/d).$$

2.4. Numerical–analytical method

It is easy to see that the analysis of composite 'triangular' solutions was purely local in nature, so it is quite possible to easily generalize them to the 'polygonal' case when the initial functions $E_0(\rho)$ and $V_0(\rho)$ have the form of continuous broken lines consisting of a finite number of straight lines segments and satisfying the boundary conditions (2.3). So, we define an arbitrary grid on the segment $[-d, d]$

$$-d = \rho_0 < \rho_1 < ... < \rho_M = d$$

and on it a set of standard piecewise linear functions (see [22])

$$
\varphi_i(\rho) = \begin{cases}
\dfrac{\rho - \rho_{i-1}}{\rho_i - \rho_{i-1}} & \text{at} \quad \rho_{i-1} \leq \rho \leq \rho_i, \\[3mm]
\dfrac{\rho_{i+1} - \rho}{\rho_{i+1} - \rho_i} & \text{at} \quad \rho_i \leq \rho \leq \rho_{i+1}, \\[3mm]
0 & \text{at other} \quad \rho
\end{cases}
$$

for $i = 1, \ldots, M - 1$. Let the initial functions have the form

$$
E_0(\rho) = \sum_{i=1}^{M-1} E_i \varphi_i(\rho), \quad V_0(\rho) = \sum_{i=1}^{M-1} V_i \varphi_i(\rho), \tag{2.29}
$$

where the numerical values of E_i, V_i are related to local gradients on the interval $[\rho_{i-1}, \rho_i]$, $i = 1, \ldots, M$ by the relations

$$
\alpha_i = \frac{E_i - E_{i-1}}{\rho_i - \rho_{i-1}}, \quad \beta_i = \frac{V_i - V_{i-1}}{\rho_i - \rho_{i-1}}. \tag{2.30}
$$

Further, we assume that for any value of i the set of parameters (α_i, β_i) satisfies the inequality $1 - 2\alpha_i - \beta_i^2 > 0$, then the corresponding axial solutions with the functions $D_i(\theta)$, $W_i(\theta)$, 'pasted together' by continuity at the points kinks are regular 'polygonal' solutions. Note that the number of kinks does not increase over time.

We present a formal sequence of actions for constructing solutions of the type under discussion.

1. Using the initial functions (2.29), we define the starting sets (2.30) for calculating axial solutions in accordance with formula (2.17). As a result, we obtain analytical solutions $D_i(\theta)$, $W_i(\theta)$, $i = 1, \ldots, M$.

2. Using the initial function $E_0(\rho)$ from (2.29), in accordance with formula (2.6), we calculate the Lagrangian coordinates of the break points ρ_i^L. The dynamics of their Eulerian coordinates in time is described by the relation (2.23) with the formal replacement of the subscript I by i. From here we have the analytical dependences $\rho_i(\theta)$, $i = 1, \ldots, M - 1$.

3. The final formulas for the analytical solution of the basic system (2.1) with the initial conditions (2.29) on the interval $[\rho_{i-1}, \rho_i]$, $i = 1, \ldots, M$, have the form

$$E(\rho,\theta)=E_{i-1}(\theta)\frac{\rho_i-\rho}{\rho_i-\rho_{i-1}}+E_i(\theta)\frac{\rho-\rho_{i-1}}{\rho_i-\rho_{i-1}},$$

$$V(\rho,\theta)=V_{i-1}(\theta)\frac{\rho_i-\rho}{\rho_i-\rho_{i-1}}+V_i(\theta)\frac{\rho-\rho_{i-1}}{\rho_i-\rho_{i-1}},$$

where the time-dependent coefficients are related to the axial solutions by the dependences

$$E_i(\theta)=E_{i-1}(\theta)+D_i(\theta)(\rho_i-\rho_{i-1}),\quad V_i(\theta)=V_{i-1}(\theta)+W_i(\theta)(\rho_i-\rho_{i-1}).$$

Recall that the solutions constructed satisfy the boundary conditions (2.3); therefore, the following relation is valid:

$$E_0(\theta)=E_M(\theta)=V_0(\theta)=V_M(\theta)\equiv0.$$

The proposed construction allows an alternative description in terms of the Lagrangian variables. Its essence lies in the integration of the equations of motion of particles (2.5), which at the initial time point are located at the break points ρ_i^L, followed by a linear continuation of the desired functions $E(\rho,\theta)$ and $V(\rho,\theta)$ between the Eulerian coordinates of these particles. We explain this in more detail.

1. Using the initial function $E_0(\rho)$ from (2.29), in accordance with formula (2.6), we calculate the Lagrangian coordinates of the break points ρ_i^L. At these points at the initial moment of time there are some particles; fix them. Since in the plane case $R(\rho^L, \theta) = E(\rho, \theta)$ for an arbitrary time, for the indicated particles from functions (2.29) we obtain the initial conditions for integrating the equations of motion (2.5).

2. The time dynamics of the Eulerian coordinates of fixed particles is described by the relation (2.23) with the formal replacement of the subscript I by i. Thus, at an arbitrary time, we have analytical dependences both for the points $\rho_i(\theta)$, $i = 1, \ldots, M - 1$, and for the values $E(\rho, \theta)$ and $V(\rho, \theta)$ at these points (see formulas (2.20)).

3. Now on each sub-interval $[\rho_i(\theta), \rho_{i+1}(\theta)]$, $i = 0, \ldots, M - 1$, where the extreme values coincide with the boundaries of the original segment $[-d, d]$, as a result of the linear interpolations we obtain $E(\rho, \theta)$ and $V(\rho, \theta)$ for arbitrary values of ρ from each sub-interval at any time θ.

The above means that the movement of particles associated with the breakpoints of the initial functions and the values of velocity and electric field at the nodes of the variable grid associated with the Eulerian coordinates of these particles completely determine the desired solution.

Note that the proposed analytical method for constructing exact 'polygonal' solutions is in fact an approximate numerical–analytical method for solving the main system in the Eulerian coordinates in the case when the initial functions $V_0(\rho)$ and $E_0(\rho)$ have sufficient smoothness. Indeed, in this case, 'polygonal' initial conditions are naturally perceived either as piecewise linear interpolants of the initial conditions, or as best approximations in the corresponding finite-dimensional metric spaces with piecewise linear bases. However, it should be borne in mind that from the point of view of computational efficiency, such an approach can be inferior to direct integration in Lagrangian variables, since the relations (2.4), (2.5) are independent for different ρ^L and, therefore, admit a parallel numerical implementation. In our case, the increase in computational work can be compensated by obtaining additional information about the spatial dependence of the desired functions. Nevertheless, the use of axial solutions should be considered, first of all, as some expansion of the analytical apparatus of research. In particular, the idea of 'closing a computational algorithm' (see [52]) (in this case, the numerical-analytical method) gives rise to the following hypothesis.

Hypothesis. *Let the initial functions $V_0(\rho)$ and $E_0(\rho)$ be continuously differentiable on the segment $[-d, d]$. Then for the existence and uniqueness of the solution $V(\rho, \theta)$, $E(\rho, \theta)$ of problem (2.1)–(2.3) continuously differentiable with respect to both variables ρ and θ of a 2π-periodic in time $\forall \theta > 0$, it is necessary and sufficient to satisfy the inequality*

$$1 - 2\frac{\partial E_0(\rho)}{\partial \rho} - \left(\frac{\partial V_0(\rho)}{\partial \rho}\right)^2 > 0$$

at each point $\rho \in (-d, d)$.

Additionally, we note that the 'polygonal' solutions are good tests for approximate algorithms for calculating the problem (2.1) – (2.3) in the Eulerian variables. At the same time, the corresponding numerical–analytical method is essentially a shock-fitting method of 'distinguishing weak discontinuities', and therefore, as a natural addition to it, it is desirable to have computationally more convenient 'shock capturing' schemes (here we use the terminology from [69]). Justified schemes of this type are currently unknown for the considered statement, but their relevance (especially in terms of generalization to the case of a larger number of independent variables) raises no doubts.

2.5. Bibliography and comments

In the chapter for the well-known quasilinear system of hyperbolic equations the so-called axial solutions are studied, i.e., solutions of some auxiliary system of ordinary differential equations. In particular, the necessary and sufficient condition for the global (2π-periodic) existence in time of solutions is obtained, and when it is executed, explicit analytical formulas are written out for the indicated solutions. On the basis of axial solutions, exact solutions of the initial system of equations for initial functions having the form of triangles are constructed. Further, the 'triangular' solutions are generalized to the case when the initial functions have the form of continuous polygonal lines consisting of a finite number of straight-line segments. And, finally, on the basis of such 'polygonal' solutions, a numerical–analytical method is proposed for solving a quasilinear system in the Eulerian variables. The results of the chapter were obtained in [101, 102].

The simplest theorem of the qualitative theory of differential equations, generating conditions under which it makes sense to raise the question of the behaviour of solutions of a differential equation on an infinite time interval, was proved in [13]. We give its wording.

Let M be a smooth (class C^r, $r \geq 2$) manifold, $\mathbf{v}$: $M \to TM$ – be a vector field. Let the vector $\mathbf{v}(\mathbf{x})$ be different from the zero vector $T_x M$ only in the compact part K of the manifold M. Then there exists a one-parameter diffeomorphism group g^t: $M \to M$, for which the field $\mathbf{v}$ is the phase velocity field:

$$\frac{d}{dt} g^t \, \mathbf{x} = \mathbf{v}\left(g^t \, \mathbf{x}\right).$$

The consequence of this theorem is the statement:
    Any solution to a differential equation

$$\dot{\mathbf{x}} = \mathbf{v}(\mathbf{x}), \quad \mathbf{x} \in M$$

can be continued back and forth indefinitely. In this case, the value *of the solution g^{t}x at time t depends smoothly on t and on the initial condition* **x**.

The compactness condition in the theorem cannot be dropped, since, for example, for $M = \mathbb{R}$, $\dot{x} = x^2$, solutions cannot be continued indefinitely.

An important tool for studying systems of differential equations of small dimension is the Poincaré–Bendixson theory [65, 66]. Recall its main position, following [66].

Let $f = (f_1, f_2)$ be a real continuous vector function defined on a bounded open subset D of a real (x_1, x_2)-plane. Consider a two-dimensional autonomous system

$$x_1' = f_1(x_1, x_2), \quad x_2' = f_2(x_1, x_2).$$

It is assumed that for each point $(\xi, \eta) \in D$ and each real t_0 there is a unique solution vector $(\varphi_1(t), \varphi_2(t))$ passing through the specified point at the time t_0. The point of the domain D at which both functions f_1 and f_2 vanish is called singular. Point D, which is not singular, is called a regular point.

Suppose that C^+ (or C^-) is the semi-infinite trajectory of the system under discussion, which represents the solution defined for all $t \geq t_0$ (or $t \leq t_0$) for some t_0. In other words, C^+ (or C^-) is the set of all points $P(t)$ of the set D with coordinates $(\varphi_1(t), \varphi_2(t))$, where $t_0 \leq t < +\infty$ (or $-\infty < t \leq t_0$). A point of a $Q(x_1, x_2)$-plane is called a limit point for C^+ (or C^-) if there is a sequence of real numbers $\{t_n\}$, $n = 1, 2, \ldots$, where $t_n \to +\infty$ (or $t_n \to -\infty$) as $n \to \infty$, such that $P(t_n) \to Q$ as $n \to \infty$. The set of all limit points of the C^+ (or C^-) semitrajectory is denoted by $L(C^+)$ (or $L(C^-)$), and these sets are called *limit sets*.

The fundamental Poincaré–Bendixson theorem says:

Let C^+ be a positive semi-infinite trajectory contained in a closed subset K of the set D. If the set $L(C^+)$ consists only of regular points, then either

(i) *the semi-infinite trajectory $C^+(= L(C^+))$ is a periodic trajectory,*

or

(ii) *the set $L(C^+)$ is a periodic trajectory.*

The established view on collapsing ('blow-up') solutions of autonomous quadratic systems of differential equations is described in the book [142]. The work [17] is devoted to non-autonomous quadratic systems. Necessary and sufficient conditions for destruction in a finite period of time for solving a system of differential equations with the right-hand side of a polynomial type were studied in [137]. It should be noted that the systems describing oscillatory processes, in the works of the 'blow-up' subjects have been focused on obviously insufficiently.

Analysis of plane non-relativistic plasma oscillations of large amplitude was apparently first touched upon in an important paper [131] and later a chapter in a monograph [129] was devoted to it. Then a cycle of works followed, one way or another based on solutions of a particular type of the P1NE system [106, 161, 163–165]. The papers [141, 169] are devoted to generalizations of these equations with an allowance for dissipation and resistance in a plasma.

It should be borne in mind that the use of axial solutions in no way contradicts the main method of studying hyperbolic systems, i.e. the method of characteristics; the proposed approach seems to be a useful and effective addition to the traditional scheme of research. In addition, we repeat about the relevance of the 'shock-capturing' schemes in the Eulerian variables for solutions that allow weak discontinuities. It seems that they are interesting from a mathematical point of view, especially in the case of several spatial variables. We emphasize that the methods for calculating plasma oscillations have been known for a long time (see, for example, [39]), but they are based either on using only the Lagrangian variables (which is very difficult to transfer to a higher dimension), or generally on discrete plasma models and/or corresponding equations for characteristics.

Simple triangular solutions of system (2.1) have already been mentioned in the literature (see [106]), true for the particular case when the initial velocity function $V_0(\rho)$ is identically zero and the signs of the derivatives of the initial function are fixed $E_0(\rho)$: $\alpha_1 < 0$, $\alpha_2 > 0$. The indicated restriction on the initial distribution of velocities leads to the fact that the initial phases of oscillations of all particles in the Lagrangian variables coincide, which causes a strong simplification of the formulas. For example, the equation of motion of a breakpoint in the Eulerian coordinates takes the following form:

$$\rho_I(\theta) = \rho_I^0 - (\rho_I^0 + d)(1 - \cos\theta)\alpha_1,$$

and the solution $E(\rho, \theta)$ itself is conveniently represented through the time-dependent factors in (2.22), if we use the special case of formulas (2.19) for $\beta = 0$:

$$D_i(\theta) = \frac{\alpha_i \cos\theta}{1 - \alpha_i(1 - \cos\theta)}, \quad i = 1, 2.$$

Note also that the existence condition for the solution found for the initial function $V_0(\rho) \equiv 0$ was given in [106]. This condition has the form $\alpha_2 < 1/2$ and is a special case of the extremely exact condition (2.15).

The condition of regularity of the motion of the Lagrangian particles (i.e., preservation of their order)

$$\frac{\partial R(\rho^L, \theta)}{\partial \rho^L} > -1$$

has already been encountered in the literature (see [131]), and it was noted there that the inequality $T(\rho^L, \theta) < 1$ is sufficient to conserve the regularity of motion in time. However, the relation between the total energy constraint and the spatial gradients of the functions $E_0(\rho)$ and $V_0(\rho)$ was not traced in the cited paper.

A special case of a 'polygonal' solution for the value $M = 3$ was presented in [106] with the following parameters:

$$\rho_1 = -\rho_2 = -d/2, \quad \alpha_1 = -\alpha_2 = \alpha_3 = \Delta, \quad \beta_1 = \beta_2 = \beta_3 = 0.$$

Analysis of its existence led to the condition $|\Delta| < 1/2$, which agrees well with the necessary and sufficient condition (2.15).

3

Plane one-dimensional relativistic electron oscillations

The chapter deals with plane one-dimensional relativistic electron oscillations described by the P1RE system of equations. First, it explains the background of breaking: the shift of the fundamental oscillation frequency and the violation of the property of invariance. Then, a basic numerical algorithm is built on the basis of the 'leapfrog' scheme and the Lagrangian variables, with the help of which the development and completion of oscillations is determined by numerical simulation. Further, in order to study more complex problems (for the future!), an approximate method is constructed based on the Eulerian variables and the idea of splitting along physical processes. Finally, using asymptotic formulas describing the dynamics of relativistic oscillations, the artificial boundary conditions of various orders of accuracy with respect to amplitude and grid parameters are derived.

3.1. Problem statement in the Eulerian and Lagrangian variables

We study one of the simplest formulations for the P1RE equations, in which a breaking effect is observed. The equations themselves, which describe free plane one-dimensional relativistic oscillatory motions of electrons, were obtained in Chapter 1 from the base system (1.5)–(1.12). Recall them:

$$\frac{\partial P}{\partial \theta} + E + V\frac{\partial P}{\partial \rho} = 0, \quad \frac{\partial E}{\partial \theta} - V + V\frac{\partial E}{\partial \rho} = 0, \quad V = \frac{P}{\sqrt{1+P^2}}. \tag{3.1}$$

The P1RE equations which take the relativistic effect into account fundamentally differ from the non-relativistic P1NE equations, considered in the previous chapter, only by the algebraic relation connecting the velocity V and the momentum P of electrons. However, the algebraic connections between the desired variables do not change the structure of the initial and boundary conditions determined solely by the properties of the differential equations. Therefore, the necessary initial and boundary conditions for the system (3.1) can be defined similarly to the non-relativistic case.

We assume that we have initial conditions

$$P(\rho,0) = P_0(\rho), \quad E(\rho,0) = E_0(\rho), \tag{3.2}$$

in the form

$$P_0(\rho) = \beta\rho\exp^2\left\{-\frac{\rho^2}{\rho_*^2}\right\}, \quad E_0(\rho) = \alpha\rho\exp^2\left\{-\frac{\rho^2}{\rho_*^2}\right\} \tag{3.3}$$

with some constants α, β. As a rule, $\beta = 0$.

The quasilinear system of equations (3.1) is the basis for the research carried out in the chapter, therefore, in addition to writing in the Eulerian variables, its form in the Lagrangian variables will be useful:

$$\frac{dP(\rho^L,\theta)}{d\theta} = -E(\rho^L,\theta), \quad \frac{dE(\rho^L,\theta)}{d\theta} = V(\rho^L,\theta), \quad V = \frac{P(\rho^L,\theta)}{\sqrt{1+P^2(\rho^L,\theta)}},$$

where $d/d\theta = \partial/\partial\theta + V\,\partial/\partial\rho$ is the total time derivative.

Recall that the function $R(\rho^L,\theta)$, which determines the displacement of a particle with the Lagrangian coordinate ρ^L:

$$\rho(\rho^L,\theta) = \rho^L + R(\rho^L,\theta), \tag{3.4}$$

satisfies the equation

$$\frac{dR(\rho^L,\theta)}{d\theta} = V(\rho^L,\theta).$$

From this it follows that the values of $R(\rho^L,\theta)$ and $E(\rho^L,\theta)$ coincide up to a constant, which is calculated from the condition that the electric field is zero in the absence of displacements. In other words, in the case of plane one-dimensional oscillations, the relation

$$R(\rho^L, \theta) \equiv E(\rho, \theta), \tag{3.5}$$

holds and the basic system of equations (3.1) in the Lagrangian variables takes the form

$$\frac{dP}{d\theta} = -R, \quad \frac{dR}{d\theta} = \frac{P}{\sqrt{1+P^2}} \equiv V. \tag{3.6}$$

We note that relation (3.4) is very useful for determining the Lagrangian coordinate of a particle ρ^L from the initial distribution $E_0(\rho)$; at the same time, the formula is as follows:

$$\rho^L = \rho(\rho^L, 0) - E_0(\rho(\rho^L, 0)). \tag{3.7}$$

To summarize the above, the trajectories of all particles, each of which is identified by the Lagrangian coordinate ρ^L, can be determined by independent integration of the system of ordinary differential equations (3.6). For this, two initial conditions are required: $R(\rho^L, 0)$ and $P(\rho^L, 0)$. From (3.3) we have $P(\rho^L, 0) = 0$. To determine $R(\rho^L, 0)$, we must first set the position of the particle ρ at the initial moment of time, then the displacement at this point is given by the electric field $R(\rho^L, 0) = E_0(\rho)$ according to the formula (3.3). In turn, the Lagrangian coordinate of the particle is determined by the formula (3.7). Knowledge of the Lagrangian coordinate ρ^L and the displacement function $R(\rho^L, \theta)$ uniquely characterizes the particle trajectory by formula (3.4).

3.2. Theoretical background of breaking

The section considers two important aspects of the investigated plasma oscillations:

— the asymptotic analysis in the weakly non-linear approximation establishes that the oscillation frequency shift quadratically depends on the amplitude of the oscillations (this is guaranteed to lead to the intersection of the trajectories of neighbouring particles, see section 1.1),

— violation in the relativistic case of the properties of the invariance of the electron density relative to the linear replacement of dependent and independent variables makes the observation of the breaking very convenient and intuitive.

3.2.1. Quadratic frequency shift

If we assume that the oscillation amplitude is sufficiently small, that is, in the initial condition (3.3) for the electric field we have $a_* \ll \rho_*$, then equations (3.1) become weakly nonlinear and their approximate solutions can be constructed using the perturbation theory method [25], [154]. We give here a brief derivation of the corresponding analytical formulas.

Given the approximate representation of the relativistic dependence of the momentum on the velocity

$$P \approx V\left(1+\frac{V^2}{2}\right)$$

and excluding the displacement R from the system (3.6), we obtain the following equation for the velocity V:

$$\left(\frac{d^2}{d\theta^2}+1\right)V+\frac{1}{2}\frac{d^2}{d\theta^2}V^3=0. \tag{3.8}$$

We add to it the initial conditions from (3.3):

$$V(\rho,0)=0, \quad \frac{dV}{d\theta}(\rho,0)=A(\rho). \tag{3.9}$$

We assume that the amplitude of the oscillations $A(\rho)$ is small, then the approximate (asymptotic) solution (3.8), (3.9) is uniformly bounded in the variable θ and differs from the exact one by the third order of smallness, i.e. $O(A^3(\rho))$. In this case, you can put

$$A(\rho) \approx -E_0(\rho)=-\alpha\rho\exp\left\{-2\frac{\rho^2}{\rho_*^2}\right\}, \quad \alpha=\left(\frac{a_*}{\rho_*}\right)^2. \tag{3.10}$$

Since the variable ρ is included in the problem (3.8), (3.9) as a parameter, it is convenient to first consider the model problem with one independent variable θ:

$$U'' + U + \frac{1}{2}\left(U^3\right)'' = 0, \quad U(0) = 0, \ U'(0) = \varepsilon \ll 1. \tag{3.11}$$

Substituting U into (3.11) as

$$U(\theta) = \varepsilon U_1(\theta) + \varepsilon^2 U_2(\theta) + \varepsilon^3 U_3(\theta) + \ldots$$

(expansion in powers of the small parameter), we find that the components $U_i(\theta)$, $i = 1, 2, 3$, are solutions of the following auxiliary problems:

$$
\begin{aligned}
U_1'' + U_1 &= 0, & U_1(0) &= 0, \quad U_1'(0) = 1, \\
U_2'' + U_2 &= 0, & U_2(0) &= 0, \quad U_2'(0) = 0, \\
U_3'' + U_3 + \frac{1}{2}\left(U_1^3\right)'' &= 0, & U_3(0) &= 0, \quad U_3'(0) = 0.
\end{aligned}
\tag{3.12}
$$

It is easy to get direct decomposition

$$U(\theta) = \varepsilon \sin\theta + \varepsilon^3 \left(-\frac{27}{64}\sin\theta + \frac{9}{64}\sin 3\theta - \frac{3}{16}\theta \sin\theta\right) + o(\varepsilon^3).$$

Note that the asymptotic solution found is not applicable for long time intervals, since its component $U_3(\theta)$ contains a growing term of the form $\theta \sin \theta$.

It is easy to get rid of the resonance term in the third equation (3.12), which generates an unlimited particular solution, following [25]. It is enough to replace the function $\sin \theta$ with $-\sin (1 + \varepsilon^2\omega_2)$ θ in U_1, i.e., to the fundamental oscillation frequency equal to one add a square-small value of ε in the parameter. Here we mean that ω_2 does not depend on ε.

For $U_1 = \sin (1 + \varepsilon^2\omega_2)\theta$, we have $U_1'' + U_1 = -2\varepsilon^2\omega_2 U_1 + O(\varepsilon^4)$. The obtained right-hand side generates an additional term of order ε^3, since the component $U_1(\theta)$ has a factor ε in the solution. By virtue of this, the last equation in (3.12) takes the form

$$U_3'' + U_3 - 2\omega_2 U_1 + \frac{1}{2}\left(U_1^3\right)'' = 0.$$

Hence, given the equality

$$\cos^3 y = \frac{3}{4}\cos y + \frac{1}{4}\cos 3y, \quad y = \frac{\pi}{2} - (1 + \varepsilon^2 \omega_2)\theta$$

and zeroing the multiplier at U_1, we obtain the correction to the fundamental frequency:

$$\omega_2 = -\frac{3}{16}. \tag{3.13}$$

Thus, the desired bounded solution of the auxiliary problem (3.11) has the form

$$U(\theta) = \varepsilon \sin \omega\theta + \varepsilon^3 \left[-\frac{27}{64}\sin \omega\theta + \frac{9}{64}\sin 3\omega\theta \right] + o(\varepsilon^3), \quad \omega = 1 - \frac{3\varepsilon^2}{16}.$$

Returning to the formulation (3.8), (3.9), we obtain the dependence of the oscillation frequency on the spatial distribution of the initial amplitude $A(\rho)$:

$$V(\rho,\theta) = A(\rho)\sin\left(1 - \frac{3A^2(\rho)}{16}\right)\theta + O(A^3(\rho)).$$

In this case, the third-order terms of smallness are no longer growing in time.

We apply the result obtained to relations (3.6), (3.7). In particular, we obtain that the trajectory of the particle, up to third-order terms of smallness, has the form

$$\rho = \rho_0 - A_0 \cos\left(1 - \frac{3A_0^2}{16}\right)\theta, \quad A_0 = A(\rho_0), \tag{3.14}$$

where ρ_0 is the Lagrangian coordinate of the particle in an equilibrium position that does not lead to the appearance of an electric field. Since the initial amplitude of oscillations $A(\rho)$ is not constant, the trajectories of some neighbouring particles must sooner or later intersect based on the considerations from Section 1.1.

3.2.2. Violation of the property of invariance

We first consider equations (3.1), neglecting the relativistic effect (P1NE-equations):

$$\frac{\partial V}{\partial \theta} + E + V\frac{\partial V}{\partial \rho} = 0, \quad \frac{\partial E}{\partial \theta} - V + V\frac{\partial E}{\partial \rho} = 0. \tag{3.15}$$

The expression (1.23) for the electron density N remains unchanged:

$$N = 1 - \frac{\partial E}{\partial \rho}.$$

We write the solution (3.15) in the form

$$V = \sigma U, \quad E = \sigma G,$$

where σ is a real parameter. If in this case we make the replacement of the independent variable $\rho = \sigma x$, then it is easy to verify that the new functions U and G still satisfy equations (3.15), but already in the variables θ and x. In addition, if in the initial conditions (3.3) with the same transformation the coordinates ρ replace the values a_* and ρ_* by σa_* and $\sigma \rho_*$, respectively, we will have the equality $E_0(\rho) = \sigma G(x, 0) \equiv G_0(x)$.

The above means a specific property of invariance: a simultaneous proportional change in the independent variable ρ and the parameters a_*, ρ_* leads to a similar change in the functions V and E; on the other hand, the noted changes $\rho = \sigma x$, $E = \sigma G$ do not change the values of electron density N in the variables θ and x, which follows from formula (1.23). This, in particular, implies that due to the selection of the parameters of the problem a_* and ρ_*, the amplitudes fixed in density can be reached at arbitrarily small (or arbitrarily large!) amplitudes of the oscillations of velocity and electric field.

It should be noted that this property is directly related to the global in time existence of plane non-relativistic oscillations [102], i.e., with the absence of a shift of the fundamental frequency depending on the amplitude. In turn, *the basic equations (3.1), studied in this chapter, do not possess this property of invariance*: this is prevented by the relation $V = P/\sqrt{1 + P^2}$, which just characterizes the relativistic effect.

Summarizing the results of Section 3.2, we formulate the main methodological idea underlying the study of plane relativistic oscillations of this chapter. First, some parameters of the problem a_*, ρ_*, $d = 4,5\rho_*$ (boundary of the computational domain) are fixed, and this set is assigned the value of the auxiliary parameter $\sigma = 1$. Without the loss of generality, we can assume that the bounded electron density function $N(\rho, \theta)$ for the specified set, due to the frequency shift, exists only on a certain interval $0 < \theta < \theta_{wb}(\sigma)$. In order to increase the influence of the relativistic effect, that is, to reduce the breaking time of the oscillations θ_{wb}, one should carry out simultaneous changes in the data of the problem:

$$a_* \to \sigma a_*, \quad \rho_* \to \sigma\rho_*, \quad d \to \sigma d, \quad \sigma > 1.$$

In turn, the values $\sigma < 1$ mean a decrease in the velocity of oscillations, that is, a weakening of the relativistic effect. For clarity, it is necessary to observe the electron density function at the initial scale with respect to the variable ρ, that is, without taking into account σ, since the proximity of graphs for different values of the parameter σ will indicate a weak influence of relativism. In other words, to study relativistic breaking, it is quite sufficient to observe the dependence of the electron density on one parameter, which characterizes (albeit indirectly) the velocity of oscillations.

3.3. Method in Lagrangian variables

This section is devoted to the numerical integration of equations (3.6) with the initial conditions (3.3), (3.7). For completeness, we present the calculation formulas of the method, the main purpose of which is to simulate the process of breaking of oscillations. Considering the symmetry (oddness) of the solution with respect to the straight line $\rho = 0$, we describe the algorithm for the segment $[0, d]$.

Suppose that in the initial moment of time $\theta = 0$, the particle with the number k is characterized by the initial position along the radius $\rho_0(k)$ and the initial displacement $R(k, 0)$, where $1 \leq k \leq M$, M is the total number of particles. The initial positions of all particles form the electric field, which has the form (3.3). On the other hand, the deviation of particles at the initial moment of time creates at the point with the coordinate $\rho_k = \rho_0(k) + R(k, 0)$ an electric field on the basis of formula (3.5). Comparing the expressions (3.3) and (3.5), we can determine the desired values of $\rho_0(k)$ and $R(k, 0)$. To do this, we define the initial spatial grid $\rho_k = kh$, where h is the discretization parameter with respect to the radial variable, which characterizes the proximity of the neighbouring particles. At the grid nodes, using formula (3.3), we calculate the values of the electric field $E_0(\rho_k)$. This electric field is formed by the displacement of particles, that is, on the basis of (3.5) we have the equations for determining the initial positions $\rho_0(k)$:

$$\rho_k = \rho_0(k) + R(k,0) \equiv \rho_0(k) + E_0(\rho_k).$$

Thus, to calculate the trajectory of each particle we obtain the initial data $\rho_0(k)$ and $R(k, 0)$ to which we must add the condition of particle immobility at the initial time from (3.3), i.e. $P(k, 0) = 0$.

Equations (3.6), as noted above, are ordinary differential equations. Therefore, they can be integrated in a numerically usual way [20], for example, according to the second order accuracy scheme traditional for the equations of motion (the so-called 'leapfrog' scheme) [139]. Let τ be a time discretization parameter, i.e. $\theta_j = j\tau$, $j \geq 0$, then the calculation formulas will have the following form:

$$\frac{P(k,\theta_{j+1/2}) - P(k,\theta_{j-1/2})}{\tau} = -R(k,\theta_j),$$

$$\frac{R(k,\theta_{j+1}) - R(k,\theta_j)}{\tau} = \frac{P(k,\theta_{j+1/2})}{\sqrt{1 + P^2(k,\theta_{j+1/2})}}. \qquad (3.16)$$

In this case, at an arbitrary time θ_j, the Euler grid variable can be calculated from the formula

$$\rho_k = \rho_0(k) + R(k,\theta_j), \quad 1 \leq k \leq M, \qquad (3.17)$$

in the nodes of which, in accordance with (3.5), the values of the electric field $E(\rho_k, \theta_j) = R(k, \theta_j)$ are determined. This technique was used to represent the electron density for illustrative purposes: the calculations used the formula of numerical differentiation of the second order of accuracy in the middle of the sub-segments:

$$N\left(\frac{\rho_{k+1} + \rho_k}{2}, \theta_j\right) = 1 - \frac{E(\rho_{k+1}, \theta_j) - E(\rho_k, \theta_j)}{\rho_{k+1} - \rho_k}. \qquad (3.18)$$

It was convenient to assume that at any moment in time on the straight line $\rho = 0$ there is a particle with the number $k = 0$, which always has no displacement, i.e. its trajectory just coincides with the axis $\rho = 0$ and on this trajectory the electric field always equals zero. Similar considerations were used for the boundary of the region $\rho = d$ and particles with the number $k = M$.

3.4. Scenario of development and completion of oscillations

For definiteness, let us fix the values of the parameters in (3.3): $a_* = 2.07$, $\rho_* = 3.0$, and we assign for them the value of the auxiliary parameter $\sigma = 1$ for further analysis, and consider the corresponding Fig. 3.1. The dotted line shows the spatial distribution of the electron density N at the initial moment of time, that is, the consequence of the formulas (1.23) and (3.3). An excess of positive charge at the origin of coordinates leads to the movement of electrons in the direction of the centre of the region, which, after half a period of oscillation, generates another distribution of the density function, also shown in Fig. 3.1 (solid line). Note that the concentration of electrons in the centre of the region can be many times greater than the equilibrium (background) value equal to unity. The fixed parameters lead to small intensity fluctuations, when the amplitude of the oscillations is only about 10 times higher than the background value. If nonlinear plasma oscillations retained their spatial form in time, then presented in Fig. 3.1 images of electron density would regularly change each other through each half of the period,

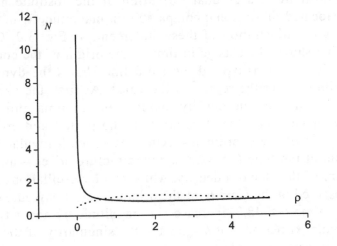

Fig. 3.1. Spatial distributions of the electron density varying through each half of the period with regular oscillations: maximum at the origin of coordinates (solid line), minimum at the origin of coordinates (dashed line).

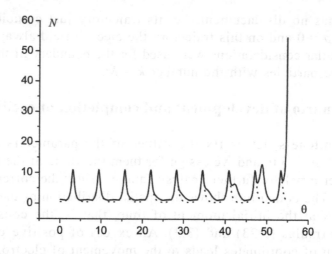

Fig. 3.2. Dynamics of electron density at $\sigma = 1$: maximum over the region (solid line) and at the origin of coordinates (dashed line).

generating in the centre of the region a strictly periodic sequence of extrema with constant amplitudes.

However, in accordance with the results of section 3.2.1, two trends are observed during the oscillation process. The first of these is that off-axis oscillations are slightly ahead in phase of the density oscillations on the axis of symmetry (with $\rho = 0$) and this phase shift increases from period to period. The second trend is more obvious: over time, there is a gradual formation of the absolute maximum density, located off-axis and comparable in magnitude with the axial ones. A good illustration of these statements is Fig. 3.2. On it the dotted line shows the change in time at the origin of the coordinates for the electron density, and the solid line shows the dynamics of the maximum over the region of the value. At first, the oscillations are regular, that is, the density maxima and minima global in the region replace each other through half the period and are located at the origin of the coordinates. After the seventh regular (central) maximum at the time $\theta \approx 42.2$, a new structure arises – an off-axis maximum of the electron density, while regular oscillations continue to be observed in the vicinity of the origin. The off-axis maximum, in turn, at a time $\theta \approx 48.8$, increases in magnitude by about two times and in the next period – at $\theta \approx 55.1$ – the singularity of the electron density appears in its place.

For greater clarity, the off-axis electron density maximum is shown in Figs. 3.3 and 3.4. Figure 3.3 shows the spatial distribution of density at the time $\theta \approx 48.8$, when it was already fully formed and became comparable in absolute value to a regular (axial) maximum. The density graph in Fig. 3.3 is a consequence of the distributions of the velocity V and the electric field E shown in Fig. 3.4. Note that in the vicinity of the density maximum the velocity function tends to jump in the derivative, and the electric field function takes on a stepwise character. It is these qualitative characteristics of V and E that ensure oscillation at the moment $\theta \approx 55.1$. It is important to note that the breaking carries the character of the 'gradient catastrophe', that is, the functions V and E themselves remain limited.

We now fix the indicated value of ρ_* and characterize the variations of the described process with a change in the parameter a_*.

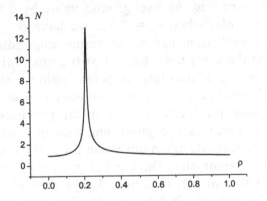

Fig. 3.3. Spatial distribution of electron density at the moment of formation of the second off-axis maximum.

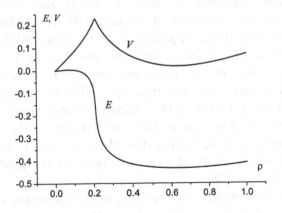

Fig. 3.4. Spatial distribution of velocity and electric field at the moment of formation of the second off-axis maximum.

Suppose first that a_* decreases monotonically. Then the evolution of plasma oscillations will stretch in time, asymptotically approaching the results of a weakly nonlinear model, which was considered in section 3.2.1. It should be noted here that, in accordance with the results of [46], for sufficiently small a_*, the asymptotic formula $\theta_{br} = C(\rho_*/a_*)^6/\rho_*^2$ with some constant C, that is, the breaking time quadratically decreases with respect to the growth of ρ_* with $a_*/\rho_* = $ const.

Now suppose that the parameter a_* increases monotonically. Then, on the contrary, the evolution of plasma oscillations will shrink with time, acquiring an increasingly nonlinear character. First of all, this will be noticeable by the absolute values of the regular axial maxima: they will begin to exceed the background value by a factor of ten or more. For example, when $a_* = 2.12$, we have $N_{axis} \approx 300$. The breaking time of oscillations and the corresponding radial coordinate will decrease. At the same time, the following qualitative picture is observed. If breaking of oscillations occur within a single period, then their radial coordinate decreases monotonically with increasing a_*. When 'skipping' the breaking time in the previous period, the radial coordinate increases in a jump, and then gradually decreases within the period, while its minimum value for periods tends to zero, i.e., to the axis of symmetry. During the growth of the parameter a_* (for a fixed value of ρ_*), it is not always possible to trace the formation and growth of the off-axis density extremum. Off-axis breaking often occurs so fast that the off-axis extremum cannot survive even one period in time. We also note that the formation time of the first axial extremum of electron density also decreases with increasing a_*. At least, not once in the calculations, the breaking was detected on the eve of the appearance of the first axial maximum. The critical value in the hydrodynamic model is $(a_*/\rho_*)^2 = 1/2$, in the vicinity of which the breaking will be almost axial: the initial distribution of electrons is such that they all rush towards the axis, are reflected from it, and then the electron trajectories very quickly intersect. Thus, the duration of the oscillations in the vicinity of the critical value will be arbitrarily close to half a period. This fact follows from the study of axial solutions of the problem under consideration [100, 102].

Let us discuss the relativistic breaking of the electronic oscillations, described in section 3.2.2. Let us leave the ratio a_*/ρ_* unchanged as the parameter ρ_* increases, that is, the characteristic scale and, accordingly, the oscillation velocity. For a quantitative description,

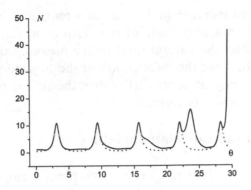

Fig. 3.5. Dynamics of electron density at $\sigma = 1.5$: maximum over the region (solid line) and at the origin of coordinates (dotted line).

we use the parameter σ: its increase, i.e. $\rho_* \to \sigma\rho_*$, corresponds to an increase in the relativistic factor. As it decreases, plane oscillations tend to be non-relativistic; they were studied in [102].

Consider Fig. 3.5, corresponding to an increase in the parameter ρ_* by 1.5 times. It should be compared with Fig. 3.2. It is easy to see that while maintaining the amplitude of oscillations in the centre of the region, the peripheral processes leading to breaking develop approximately two times faster. Thus, the off-axis extremum, comparable in magnitude with the axial one, was formed at $\theta \approx 23.7$, which led to a breaking at the time instant $\theta \approx 29.5$. This trend continues in the future. Figure 3.6 shows electron density plots corresponding to the parameter $\sigma = 2$, that is, a further increase in ρ_*. Here, the breaking time was approximately halved compared with the previous version. Thus, the influence of the relativistic factor

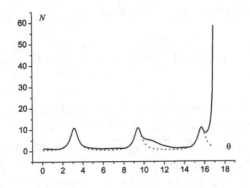

Fig. 3.6. Electron density dynamics at $\sigma = 2$: maximum over the region (solid line) and at the origin (dotted line).

associated with an increase in ρ_* leads to a reduction in the breaking time with a constant amplitude of oscillations at the centre of the region. In this case, the natural limit of the breaking time is half the oscillation period, since the intersection of the trajectories outside the axis of symmetry cannot occur earlier than the first regular maximum of the electron density is formed.

3.5. Method in the Eulerian variables

We convert the equations (3.1) to a convenient form:

$$\frac{\partial P}{\partial \theta} + E + \frac{\partial \gamma}{\partial \rho} = 0, \quad \gamma = \sqrt{1 + P^2},$$

$$V = \frac{P}{\gamma}, \quad \frac{\partial E}{\partial \theta} - V + V \frac{\partial E}{\partial \rho} = 0. \tag{3.19}$$

In the system (3.19), the interaction of two physical processes is presented: nonlinear oscillations at a fixed point in space and their space–time transfer. Therefore, as in [79], we construct a splitting scheme for physical processes using the Lax–Wendroff scheme ('tripod') for transfer of equations [7].

We relate to the description of the process of nonlinear oscillations of the equation

$$\frac{\partial \tilde{P}}{\partial \theta} + \tilde{E} = 0, \quad \tilde{V} = \frac{\tilde{P}}{\sqrt{1 + \tilde{P}^2}}, \quad \frac{\partial \tilde{E}}{\partial \theta} - \tilde{V} = 0. \tag{3.20}$$

and to their transfer in space and time –

$$\frac{\partial \bar{P}}{\partial \theta} + \frac{\partial \bar{\gamma}}{\partial \rho} = 0, \quad \bar{\gamma} = \sqrt{1 + \bar{P}^2}, \quad \bar{V} = \frac{\bar{P}}{\bar{\gamma}}, \frac{\partial \bar{E}}{\partial \theta} + \bar{V} \frac{\partial \bar{E}}{\partial \rho} = 0. \tag{3.21}$$

As a basis for time sampling of both systems, we use the usual step-by-step (leapfrog) scheme [139]. Let τ be a time step, then we will assign the values E, $\tilde{E}$, $\bar{E}$, N, to 'whole' moments of time $\theta_j = j\tau$ ($j \geq -$ integer) to 'half-integers' $\theta_{j\pm1/2} - P$, $\tilde{P}$ $\bar{P}$ and also the values of γ and V depending on the momentum P. The choice of the corresponding time moment for the function value will be denoted by superscript. For discretization in space, we will use a grid with a constant step h such that $\rho_m = mh$, $0 \le m \le M$, $Mh = d$.

We write the difference equations approximating systems (3.20) and (3.21). For the first one we get

$$\frac{\tilde{P}_m^{j+1/2} - \tilde{P}_m^{j-1/2}}{\tau} + \tilde{E}_m^j = 0, \quad \tilde{V}_m^{j+1/2} = \frac{\tilde{P}_m^{j+1/2}}{\sqrt{1 + \left(\tilde{P}_m^{j+1/2}\right)^2}},$$

$$\frac{\tilde{E}_m^{j+1} - \tilde{E}_m^j}{\tau} - \tilde{V}_m^{j+1/2} = 0, \tag{3.22}$$

$$\tilde{P}_m^{j-1/2} = P_m^{j-1/2}, \quad \tilde{E}_m^j = E_m^j, \quad 1 \le m \le M - 1.$$

Before writing down the approximation of system (3.21), we recall that time discretization in the 'tripod' scheme for a model equation (such as nonlinear transfer)

$$\frac{\partial u}{\partial t} + \frac{\partial G(u)}{\partial x} = 0, \quad G(u) = \frac{u^2}{2}$$

has the appearance

$$\frac{u^{j+1} - u^j}{\tau} + \frac{\partial G^j}{\partial x} = \frac{\tau}{2} \frac{\partial}{\partial x} \left(A^j \frac{\partial G^j}{\partial x} \right),$$

where $A = \dfrac{\partial G}{\partial u}$, and the superscript denotes the belonging of the function to the corresponding time moment. If, however, as the model equation we take the linear transport equation

$$\frac{\partial u}{\partial t} + v \frac{\partial u}{\partial x} = 0,$$

then the time discretization, analogous to the 'tripod' scheme, for it, respectively, looks like this:

$$\frac{u^{j+1} - u^j}{\tau} + \left(v^j + \frac{\tau}{2} \frac{\partial v}{\partial t} \right) \frac{\partial u}{\partial x} = \frac{\tau v^j}{2} \frac{\partial}{\partial x} \left(v^j \frac{\partial u}{\partial x} \right),$$

and also has an $O(\tau^2)$ approximation on smooth solutions.

The discrete analogue of the system (3.21), convenient for realization, built on the basis of the given model schemes, has the following form:

$$\frac{\overline{P}_m^{j+1/2} - \overline{P}_m^{j-1/2}}{\tau} + \overline{\gamma}_{\ddot{X},m}^{j-1/2} = \frac{\tau}{2}\left(\overline{V}_{s,m}^{j-1/2}\overline{\gamma}_{X,m}^{j-1/2}\right)_{\overline{X},m},$$

$$\overline{\gamma}_m^{j+1/2} = \sqrt{1+\left(\overline{P}_m^{j+1/2}\right)^2}, \quad \overline{V}_m^{j+1/2} = \frac{\overline{P}_m^{j+1/2}}{\overline{\gamma}_m^{j+1/2}},$$

$$\frac{\overline{E}_m^{j+1} - \overline{E}_m^{j}}{\tau} + \left(\overline{V}_m^{j+1/2} + \frac{\tau}{2}\frac{\overline{V}_m^{j+1/2} - \overline{V}_m^{j-1/2}}{\tau}\right)\overline{E}_{\ddot{X},m}^{j} = \tag{3.23}$$

$$= \frac{\tau}{2}\overline{V}_m^{j+1/2}\left(\overline{V}_{s,m}^{j+1/2}\overline{E}_{X,m}^{j}\right)_{\overline{X},m}$$

$$\overline{P}_m^{j-1/2} = \tilde{P}_m^{j+1/2}, \quad \overline{E}_m^{j} = \tilde{E}_m^{j+1}, \quad 1 \le m \le M-1,$$

$$\overline{P}_0^{j+1/2} = \overline{P}_M^{j+1/2} = \overline{E}_0^{j+1} = \overline{E}_M^{j+1} = 0.$$

Expression (3.23) used the following notations: $F_{\ddot{X},m} = (F_{m+1} - F_{m-1})/(2h)$ is the central difference, $F_{X,m} = (F_{m+1} - F_m)/h$ and $F_{\overline{X},m} = (F_m - F_{m-1})/h$ are the forward and backward differences, respectively, $F_{s,m} = (F_{m+1} + F_m)/2$.

After calculations using the scheme (3.23), one should redefine the required functions at the next time layer:

$$P_m^{j+1/2} = \overline{P}_m^{j+1/2}, \quad E_m^{j+1} = \overline{E}_m^{j+1}, \quad 0 \le m \le M,$$

and calculate (if necessary) the value of the electron density by the formula

$$N_m^{j+1} = \begin{cases} 1 - \dfrac{E_{m+1}^{j+1} - E_{m-1}^{j+1}}{2h} & \text{at } 1 \le m \le M-1, \\[3mm] 1 - \dfrac{E_1^{j+1}}{h} & \text{at } m = 0, \\[3mm] 1 & \text{at } m = M. \end{cases} \tag{3.24}$$

At this time, the calculations at the j-th time step end and one can proceed to the next step. It should be noted that the initial data (3.3) correspond to $j = 0$; therefore, they should be attributed to P for the layer with number $-1/2$, and for E – to number 0.

We make a remark about the proposed splitting scheme (3.22), (3.23). For each of the auxiliary problems with sufficient smoothness of the solution there is an approximation of order $O(\tau^2 + h^2)$, as well as the stability condition obtained on the basis of the spectral attribute [7, 20], of the form $\tau = O(h)$. This allows us to achieve significant savings in computational resources due to such a weak stability condition without loss of approximation. In addition, the scheme (3.22), (3.23) is explicit, which generates the potential for parallelization when generalized to multidimensional cases.

It should be noted that a single algorithm based on Lagrangian variables is sufficient for analyzing only the effect of breaking plane relativistic oscillations. Moreover, the calculated particle trajectories actually generate at each time moment a non-uniform Eulerian grid, most suitable for describing (plotting) the electron density function. In this sense, a scheme based on the Eulerian variables is useful, first of all, for controlling Lagrangian calculations. However, there are a number of more complex problems associated with plane electron oscillations, for example, taking into account the effects of ionization and recombination, viscosity, resistance and dissipativity of a plasma (see [141, 143, 161, 169]). The list should include tasks that take into account the movement of ions. For such statements, the construction of algorithms based on the Lagrangian variables becomes difficult, although the Eulerian methods allow a natural simple generalization to these cases. Therefore, we can say that the scheme in Euler variables is more focused on advanced investigations than on the current ones.

3.6. Artificial boundary conditions

From the results of section 3.2 it follows that at some distance from the point $\rho = 0$, when the oscillation amplitude $E_0(\rho)$ becomes sufficiently small (of the order of ε), the solution to the Cauchy problem (3.1), (3.3) with an accuracy of $O(\varepsilon^3)$ is described by the asymptotic formulas

$$V_a(\rho,\theta) = -E_0(\rho)\sin\left[\omega(\rho)\theta)\right] = P_a(\rho,\theta),$$

$$E_a(\rho,\theta) = E_0(\rho)\cos\left[\omega(\rho)\theta)\right], \quad \omega(\rho) = 1 - \frac{3}{16}[E_0(\rho)]^2. \tag{3.25}$$

Let us take these formulas as the basis for constructing artificial boundary conditions in order to limit the size of the computational domain. The main task is the maximum reduction in the size of the area which does not affect the effect of breaking oscillations. For convenience, we denote the computational domain by the variable ρ as $|\rho| \leq d$ and we will discuss the specification of the boundary conditions for $\rho = \pm d$. In terms of asymptotic analysis carried out in section 3.2.1, the artificial boundary conditions will have a different order with respect to the parameter ε – the oscillation amplitude.

3.6.1. Full damping of oscillations

For some d, we set zero boundary conditions

$$P(\pm d,\theta) = 0, \quad E(\pm d,\theta) = 0. \tag{3.26}$$

From a formal point of view, this approach is incorrect. The point is that the initial function $E_0(\rho)$ in (3.3) is non-zero on the whole axis OX and setting the conditions of the form (3.26) immediately forms a discontinuous electric field. This means that at $\rho = \pm d$, by virtue of formula (1.23), the electron density is singular, i.e. immediate breaking of oscillations.

But this is only the case with differential analysis. In numerical analysis (the construction of approximate solutions under the conditions of computational error), this formal obstacle can be overcome as follows. We introduce on the segment $|\rho| \leq d$ an uniform grid with a sufficiently small step h. Then, in the internal nodes of the grid $|\rho_k| < d$ we define the values of the initial function as $E_0(\rho_k)$, and in the boundary ones, where $\rho = \pm d$, we set zero values in accordance with (3.26). Next, we construct a sufficiently smooth function, using the values thus selected as interpolation. In particular, the natural cubic interpolation spline is quite suitable for this purpose. The available estimates of the proximity of an interpolating function of order $O(h^4)$ (see, for example, [22, 89]) lead to a natural restriction: $|E_0(\pm d)|$ must be in order less than the interpolation error. And, ideally, to coincide with the computational error, that is, with the rounding errors. Then the computational stability of the spline completely eliminates the discontinuity in the initial condition.

As mentioned earlier, it is easy to satisfy this constraint: it suffices to take d sufficiently large. For example, setting $d = 4.5\rho_*$, we get $\exp^2\{-d^2/\rho_*^2\} \approx 2.5768 \cdot 10^{-18}$. This means that, in double-precision calculations, the jump in the initial function E at the point $\rho = d$ is commensurate with the computer accuracy, that is, with the usual error of rounding off the data. In other words, in the numerical simulation of oscillations, the effect of their breaking at such a remote border will not be noticeable at all, which fully corresponds to the concept of an 'artificial boundary' [60].

The described approach to the construction of the artificial boundary conditions – the 'cutting off' of an infinite region with the help of homogeneous boundary conditions of the first kind – is practically quite convenient and therefore the most frequently used. It generates boundary conditions of the first order of accuracy – $O(\varepsilon)$. Comparison of the numerical solutions obtained in the Eulerian variables (using an artificial boundary) with solutions in the Lagrangian variables (without using an artificial boundary), carried out in [46, 79, 94], allowed us to conclude that the 'full damping' of the oscillations can be used in the numerical simulation of the breaking effect. However, the main disadvantage of this approach is an excessive increase in the computational domain. The desired breaking effect is usually observed in the vicinity of the beginning of the coordinate ρ at a distance of less than $0.1 \rho_*$, therefore more than 90% of the calculations are a kind of 'fee' for not knowing the appropriate boundary conditions.

3.6.2. Linearization of the original equations

From the asymptotic formulas (3.25) it clearly follows that all the desired functions have the same exponential decay with increasing $|\rho|$. This means that, starting with some $|\rho| = d$, the quadratic terms in the original equations (3.1) can be neglected, because of their smallness with respect to the linear terms. In other words, it is quite reasonable to set the boundary conditions for $\rho = \pm d$, $\theta > 0$

$$\frac{\partial P}{\partial \theta} + E = 0, \quad \frac{\partial E}{\partial \theta} - V = 0, \quad V = P. \tag{3.27}$$

In this case, the initial function $E_0(\rho)$ does not require any change. Such an approach generates second-order boundary conditions, $O(\varepsilon^2)$.

For the boundary conditions (3.27), using the illustrative version from [94] – $a_* = 2.07$, $\rho_* = 3$ – a series of computational experiments were conducted in order to determine the appropriate value of the parameter d. As usual, the main calculation was carried out for equations in the Euleria variables (3.1), and the control calculation (with any required accuracy) for the equations in the Lagrangian variables (3.6).

The calculations have shown that for accurate modelling of the breaking effect it is enough to take $d = 2.0\ \rho_*$. This means more than halving the computational costs as compared with the use of 'full damping' of oscillations with the same required accuracy of calculation. Additional convenience of the boundary conditions (3.27) lies in the fact that their implementation is a simplification of the difference equations used at the interior points of the domain, which consists in simply discarding quadratic terms.

3.6.3. Accounting for the weak nonlinearity of the original equations

Analysis of equations (3.1) in the weak nonlinearity approximation led to the asymptotic formulas (3.25). The time-uniform suitability of these formulas is based on the dependence of the oscillation frequency on their amplitude. This means that, if we explicitly take into account the change in the oscillation frequency, it is possible to achieve the acceptability of the modified boundary conditions at a shorter distance from the origin. We explain this in more detail.

Consider a particle with a Lagrangian coordinate $\xi = d$ (or a particle with a symmetric coordinate $\xi = -d$, since the reasoning for these two cases is the same). The amplitude of its oscillations is determined using (3.10) as $E_0(d)$, therefore, the frequency of its oscillations can be calculated using the formula $\omega(d) = 1 - \dfrac{3}{16}\left[E_0(d)\right]^2$. In accordance with formulas (3.25), this particle at an arbitrary time moment creates, by its displacement, an electric field in the vicinity of the point $\rho = d$. The difference between the asymptotic formula and the exact solution of the original equations (3.1) or their analogues in Lagrangian coordinates (3.6) has a third order of smallness with respect to amplitude, i.e. in terms of the asymptotic expansion – $O(\varepsilon^3)$. In this case, the deviation of the particle from the equilibrium position for the period is zero. Therefore, to construct the artificial boundary conditions of the third order of accuracy, one can use explicit formulas (3.25). However, it seems more convenient, first of all, for the numerical solution, to apply differential equations that generate solutions of the form (3.25).

In particular, it is proposed for $\rho = \pm d$, $\theta > 0$ to use the boundary conditions of the form

$$\frac{\partial P}{\partial \theta} + \omega(\pm d)E = 0, \quad \frac{\partial E}{\partial \theta} - \omega(\pm d)V = 0, \quad V = P. \qquad (3.28)$$

Note that in this case, the initial function $E_0(\rho)$, as well as in the previous case, does not require any change.

For the boundary conditions (3.28), using the parameters from [94] – $a_* = 2.07$, $\rho_* = 3$ – a series of computational experiments were conducted in order to determine the appropriate value of the parameter d. The calculations have shown that for an accurate simulation of the breaking effect, it is enough to take $d = 1.3\rho_*$. This means a reduction in computational costs of more than three times as compared with the use of 'full damping' of oscillations with the same required accuracy of calculation. The boundary conditions (3.28) are also easily realizable due to the rejection of the quadratic terms in the difference scheme used and the introduction of a correction factor for the oscillation frequency.

3.6.4. Deterioration of the approximation at the boundary

Another useful technique for constructing the artificial boundary conditions is the 'deterioration' (or simplification) of the approximation of the original equations. We explain this in more detail.

Let us take as an example for equations (3.1) the simplest scheme with time stepping (leapfrog) and approximation by central differences for spatial derivatives. On the grid $\theta_j = j\tau$, $\rho_k = kh$ we will have

$$\frac{P_k^{j+1/2} - P_k^{j-1/2}}{\tau} + E_k^j + \frac{P_{k+1}^{j-1/2} - P_{k-1}^{j-1/2}}{2h} = 0,$$

$$\frac{E_k^{j+1} - E_k^j}{\tau} - V_k^{j+1/2} + V_k^{j+1/2}\frac{E_{k+1}^j - E_{k-1}^j}{2h} = 0.$$

These difference equations have an approximation of $O(h^2)$, but cannot be used in the boundary nodes, for example, at the node $\rho_M = d$, due to the underdeterminedness of the functions used for ρ_{M+1}.

Replace the approximations of spatial derivatives in the node $\rho_M = d$ with one-sided differences:

$$\frac{P_M^{j+1/2} - P_M^{j-1/2}}{\tau} + E_M^j + \frac{P_M^{j-1/2} - P_{M-1}^{j-1/2}}{h} = 0,$$

$$\frac{E_M^{j+1} - E_M^j}{\tau} - V_M^{j+1/2} + V_M^{j+1/2} \frac{E_M^j - E_{M-1}^j}{h} = 0.$$

(3.29)

As a result, the approximation order will deteriorate to $O(h)$, but then the formulas will become suitable for calculations. Similarly, we can do for the node $\rho_{-M} = -d$.

Now let us consider the approximation constructed on the boundary (new artificial boundary conditions) (3.29) from the point of view of asymptotic analysis. Simple rejection of terms of the form $V\dfrac{\partial E}{\partial \rho}$, as in section 3.6.2, generates second-order conditions of accuracy with respect to the amplitude of oscillations, i.e. $O(\varepsilon^2)$. In this case, the terms are not discarded entirely: only one of the factors introduces an error of order $O(\varepsilon h)$. Since the second factor has order $O(\varepsilon)$, as a result of this approach, the error introduced into the original equations acquires the asymptotic behaviour $O(\varepsilon^2 h)$. Of course, the conditions (3.29) are asymptotically worse than the conditions (3.28), but they are in no way inferior to the conditions (3.27).

For the boundary conditions (3.29), using the parameters from [94] – $a_* = 2.07$, $\rho_* = 3$ – a series of computational experiments were conducted in order to determine the appropriate value of the parameter d. The calculations have shown that for an accurate simulation of the breaking effect, it is sufficient to take $1.5\,\rho_* \le d \le 1.8\,\rho_*$. This means that the calculations are completely consistent with the available theoretical (that is, asymptotic) estimates.

Of course, it should be noted that the coefficients at ρ_* in all formulas for d were determined from a numerical experiment approximately, with an error of about 10%.

At the end of the section, we formulate recommendations regarding the use of the proposed artificial boundary conditions. The discussed boundary conditions have different values for practical use; therefore, a reasonable combination of them should be chosen, taking into account the specifics of a particular setting.

The first order boundary conditions are very simple and reliable; therefore, they can serve as a basis for conducting initial experiments. However, they require an excessive volume of calculations, which should be avoided with large series of calculations.

The boundary conditions of the third order of accuracy are very economical, but practically not applicable for problems of dimensions of two or more in the spatial variables. This is due to the fact that in multidimensional formulations it is unlikely to determine analytically the oscillation frequency of the Lagrangian particles due to the complexity of the problem (see, for example, [124]).

The above means that, first of all, the second-order accuracy conditions (3.27) are recommended. A useful addition to them can be conditions of the form (3.29). Both types of conditions are easily realizable, they are easily generalized to spatially multidimensional problems, and also lead to a significant reduction of little significant calculations by reducing the computational domain. In addition, the boundary conditions (3.27) and (3.29) are non-reflective to both sides, i.e. they do not prevent the transfer of disturbances when the sign of the transfer coefficient changes (functions $V(\rho, \theta)$). Later in the book, the application of the artificial boundary conditions for the numerical simulation of two-dimensional relativistic electron oscillations in a plasma will be shown, and the conditions (3.26), (3.27) and (3.29) will be involved.

3.7. Bibliography and comments

This chapter is methodologically important for understanding the results of the monograph as a whole. Using the example of the simplest problem, it describes the use of both asymptotic and numerical methods for simulating the effect of the breaking of electron oscillations in a plasma.

First, it explains the background of breaking: the shift of the fundamental oscillation frequency and the violation of the property of invariance. Here, in the weakly non-linear approximation using the Lagrangian variables, it is established that the shift of the oscillation frequency quadratically depends on the initial amplitude.

The criterion for breaking the plasma oscillations with such a relationship was first formulated in [131] on the basis of considerations that the trajectories of two electrons, initially separated from each other by ρ by a distance equal to twice the amplitude of the displacement ($\Delta\rho_0 = 2A_0$), will subsequently intersect when the difference phases will be equal to π. In Ref. [46], it was clarified that the breaking is due to the intersection of the trajectories of two nearest particles, rather than the particles being separated by a distance equal to twice the displacement amplitude. However, the fundamental conclusion in any case is that the oscillations breaking is due to the difference in the oscillation frequencies of the neighbouring particles at any (arbitrarily small!) amplitudes.

The fact of the frequency shift depending on the amplitude with relativistic oscillations is well known. In particular, in [149], the breaking of the oscillations was associated with the dependence of frequency on amplitude. Important and new here is the result showing that the breaking occurs outside the symmetry axis of the problem, i.e. when $\rho \neq 0$. The justification for this is based on the fact that the axial solutions of the relativistic and non-relativistic equations are the same. Also new is the fact that the non-relativistic equations possess some useful property of invariance, while the relativistic ones do not. This makes it convenient to observe the process of breaking at a fixed (sufficiently large) amplitude of oscillations of the electron density in the vicinity of the axis of symmetry.

The particle dynamics in the Lagrangian variables is described by independent ordinary differential equations. The equations of particle trajectories are smooth functions of time; therefore, the choice of algorithms for numerical integration is wide enough [22]. The 'leapfrog' scheme was chosen taking into account economy and preservation of traditions. A detailed analysis of the scheme, including frequency correction, is given in [139]

The situation with the equations in the Eulerian variables (3.1) is much more complicated. They have a non-conservative form, so the use of traditional algorithms [69] presents known difficulties. In particular, as mentioned above, most of the methods are based on an exact or approximate solution of the Riemann problem (the Cauchy problem with discontinuous initial data) [69]. In the case of plasma oscillations, the very formulation of the Riemann problem is devoid of physical meaning: the discontinuous function of the electric field is an attribute of the breaking effect; at this moment, the use of the classical electrodynamics model ceases.

We also note the presence of source terms in the equations. Of course, the classical Lax–Wendroff and McCormack schemes can be generalized and modified for this case [151], however, such schemes are badly adapted for calculating gradient catastrophes, since the second order of accuracy involves, first of all, modelling smooth solutions.

In view of the above, a special difference scheme was developed for the equations in the Eulerian variables. It is focused on such problem statements associated with oscillations, for which the use of the Lagrangian variables in the future seems to be of little prospect. Such tasks are, for example, taking into account the dynamics of ions or spatially multidimensional electron oscillations. Within the framework of the constructed splitting scheme, all the results obtained using an algorithm based on the Lagrangian variables were completely quantitatively reproduced. The artificial boundary conditions of various orders of accuracy were also constructed in order to reduce the computational domain when using the Eulerian variables.

The construction of the artificial boundary conditions relies heavily on the properties of the solution to the original problem and is oriented, first and foremost, to ensure that the given conditions at the boundary have a minimal influence on the solution inside the region, and ideally have no effect whatsoever. The most comprehensive review of ideas and methods for constructing the boundary conditions, as well as a detailed bibliography on this subject, are contained in the monograph [60]. However, it was not possible to select a suitable example (analogue) in the case of plasma oscillations considered in it, therefore the material on the construction of the corresponding artificial boundary conditions was focused on filling this gap.

The results presented in this chapter were obtained in [94, 104].

We note again that the algorithm for calculating in the Lagrangian variables, described in section 3.3, is the simplest of all methods by which the relativistic breaking of oscillations can be calculated. In order to increase the reliability of the results of computational experiments, as a rule, control calculations were performed using a different model (in the Eulerian variables), as well as another method of integrating a system of ordinary differential equations (the classical Runge–Kutta method of the fourth order of accuracy [22] within the same particle model in the Lagrangian variables). All the calculations described above were fully reproduced within the framework of the Euler model, but there it required the refinement of the spatial and temporal grid steps by about 8–16 times. In turn, the integration of the equations of motion (3.6) by the Runge–Kutta method of the fourth order requires an increase in the computation volume by about four times, but does not give a significant gain in accuracy, since the 'working' integration step in time is quite small (about 10^{-3}), and the calculated trajectories are fairly smooth functions. The above allowed us to define algorithm (3.16) as the best from the point of view of the 'price–quality' criterion. Note that formula (3.18) for a visual representation of the electron density function is also in some sense the best. The point is that relation (3.17) generates an optimal non-uniform Eulerian grid characterizing the mutual arrangement of particles at an arbitrary time. It is the tracking of this characteristic that is the essence of the control for breaking the oscillations in Lagrangian variables. Accordingly, a more accurate way of representing the electron density as a function of the distance between the particles, given the deterioration of smoothness, is even theoretically difficult to imagine. In this case, of course, it should be borne in mind that the results of calculations by the formula (3.18) are not an attribute of the integration algorithm over time and are used solely for illustrative purposes.

The problem statement considered in the chapter leads to a gradient catastrophe. This is the name of the breaking effect, if we use the terms that are characteristic of the theory of quasilinear hyperbolic systems of equations. It should be noted that traditional methods of diagnosing such effects are practically inapplicable in this situation. Therefore, a special two-stage approach was proposed in [103, 121] to analyze the occurrence of a gradient catastrophe.

At the first stage, the spatial gradients of the initial functions form the initial conditions for a special system of the ordinary differential equations. This leads to the consideration of the simpler Cauchy problem, for which the condition for the existence of the vertical asymptote ('blow-up'-properties) should be determined. Since the auxiliary solution determines the local gradients of the original problem, the fulfillment of the specified condition will generate a gradient catastrophe, that is, a loss of continuity of the smooth solution. This stage is fundamentally different from other known approaches to the study of quasilinear hyperbolic systems, such as the continued system ([81], p. 32), the majorant system ([81], p. 62), the nonlinear capacity method [80], etc. We should note only a certain affinity with the ideas of [145], although the analysis proposed in [121] seems to be technically simpler. The first stage, devoted to the study of the properties of the auxiliary Cauchy problem for a system of ordinary differential equations, is in essence connected with the study of the initial formulation in the Eulerian variables.

The second stage is intended for a 'finer' analysis of the original problem in an equivalent Lagrangian formulation. It is carried out under the assumption that the initial data on the results of the first stage do not lead to a rapid gradient catastrophe, i.e., the breaking of oscillations arising in the first period. The essence of the stage is the construction of uniformly suitable asymptotic solutions of the original system in a weakly non-linear approximation. Here, the dependence of the oscillation frequency on the initial amplitude is derived for the equations describing the trajectories of individual particles. On this basis, a conclusion is drawn about the intersection of the particle trajectories, and, if necessary, it is possible to estimate the asymptotics of this event depending on the time duration of the oscillations.

4

Cylindrical one-dimensional relativistic and non-relativistic electron oscillations

The chapter deals with cylindrical one-dimensional relativistic and non-relativistic electron oscillations. The focus is on a simpler model. Regardless of the relativism, the cylindrical electron oscillations end in a rollover. For the needs of numerical simulation, in addition to the method based on Lagrangian variables, a splitting algorithm is constructed based on Eulerian variables. With the help of both approaches, a scenario of development and completion of the cylindrical oscillations is described and analyzed. In addition, material related to spherical oscillations is summarized.

4.1. Problem statements in Eulerian and Lagrangian variables

From the basic equations of the plasma model under consideration (1.5)–(1.12), we obtain a system whose solutions have axial (cylindrical) symmetry.

We will denote independent variables in a cylindrical coordinate system in the usual way – r, φ, z. Apply the assumptions that

- ions, due to the multiple excess of electrons by mass, are considered immobile;
- the solution is determined only by the r-components of the vector functions $\mathbf{p}_e$, $\mathbf{v}_e$, $\mathbf{E}$;
- there is no dependence in these functions on the variables φ and z, that is, $\partial/\partial\varphi = \partial/\partial z = 0$.

Then from system (1.5)–(1.12) non-trivial equations follow:

$$\frac{\partial n}{\partial t} + \frac{1}{r}\frac{\partial}{\partial r}(rnv_r) = 0, \quad \frac{\partial p_r}{\partial t} + v_r\frac{\partial p_r}{\partial r} = eE_r,$$

$$\gamma = \sqrt{1 + \frac{p_r^2}{m^2c^2}}, \quad v_r = \frac{p_r}{m\gamma}, \quad \frac{\partial E_r}{\partial t} = -4\pi env_r. \tag{4.1}$$

In this situation, the subscript e, which characterizes the connection of variables with the electron component of the plasma, is of low content and is therefore omitted.

We introduce dimensionless quantities

$$\rho = k_p r, \quad \theta = \omega_p t, \quad V = \frac{v_r}{c}, \quad P = \frac{p_r}{mc}, \quad E = -\frac{eE_r}{mc\omega_p}, \quad N = \frac{n}{n_0},$$

where $\omega_p = (4\pi e^2 n_0/m)^{1/2}$ is the plasma frequency, n_0 is the value of the unperturbed electron density, $k_p = \omega_p/c$. In the new variables, system (4.1) takes the form

$$\frac{\partial N}{\partial \theta} + \frac{1}{\rho}\frac{\partial}{\partial \rho}(\rho N V) = 0, \quad \frac{\partial P}{\partial \theta} + E + V\frac{\partial P}{\partial \rho} = 0,$$

$$\gamma = \sqrt{1 + P^2}, \quad V = \frac{P}{\gamma}, \quad \frac{\partial E}{\partial \theta} = NV. \tag{4.2}$$

From the first and last equations (4.2) follows

$$\frac{\partial}{\partial \theta}\left[N + \frac{1}{\rho}\frac{\partial}{\partial \rho}(\rho E)\right] = 0.$$

This relationship is true both in the absence of plasma oscillations ($N \equiv 1$, $E \equiv 0$), and in their presence. Therefore, from here we have a simpler expression for the electron density:

$$N = 1 - \frac{1}{\rho}\frac{\partial}{\partial \rho}(\rho E). \tag{4.3}$$

Now, excluding from the system (4.2) the density N and the actor γ, we arrive at the equations describing the free cylindrical one-dimensional relativistic oscillatory motion of electrons in a cold ideal plasma:

$$\frac{\partial P}{\partial \theta} + E + V\frac{\partial P}{\partial \rho} = 0, \quad \frac{\partial E}{\partial \theta} - V + \frac{V}{\rho}\frac{\partial}{\partial \rho}(\rho E) = 0, \quad V = \frac{P}{\sqrt{1+P^2}}. \quad (4.4)$$

Let us introduce the abbreviation C1RE (Cylindrical 1-dimension Relativistic Electron oscillations) for this system.

Considering the specificity of the cylindrical coordinate system ($\rho \geq 0$), we will study the solutions of the C1RE system in the $\{(\rho, \theta)$ semi-infinite-strip: $0 < \rho < < d, \theta > 0\}$. The introduction of the parameter d, limiting the spatial domain, is based on the results of previous chapters. Equation of equations (4.4) with localized in space initial conditions

$$P(\rho,0) = P_0(\rho), \quad E(\rho,0) = E_0(\rho), \quad \rho \in [0,d], \quad (4.5)$$

and the corresponding boundary conditions

$$P(0,\theta) = P(d,\theta) = E(0,\theta) = E(d,\theta) = 0 \quad \forall \theta \geqslant 0. \quad (4.6)$$

From where, in particular, using the algebraic relation connecting the velocity V and the momentum P of electrons, the equalities follow

$$V(0,\theta) = V(d,\theta) = 0 \quad \forall \theta \geqslant 0,$$

not allowing the transfer of perturbations of the unknown functions across the boundaries of the segment [0, d].

As before, we will keep in mind the following functions as initial conditions (4.5):

$$P_0(\rho) = 0, \quad E_0(\rho) = \begin{cases} \left(\dfrac{a_*}{\rho_*}\right)^2 \rho\exp^2\left\{-\dfrac{\rho^2}{\rho_*^2}\right\}, & 0 \leqslant \rho < d, \\ 0, & \rho = d. \end{cases} \quad (4.7)$$

Such a perturbation of the electric field, which initiates oscillations, as already indicated earlier, is characteristic of the passage of a short, powerful focus-focused laser pulse through the plasma, which has a Gaussian intensity distribution in space. By virtue of the exponential decay of the function $E_0(\rho)$, in order to ensure with sufficient accuracy the boundary condition (4.6), it suffices to set $d = 4.5\rho_*$.

To compare the presence/absence of effects of relativism, a simpler system of equations is required. We make an additional assumption that the velocity of electrons is essentially nonrelativistic, i.e.

$$P \approx V, \quad \frac{\partial P}{\partial \rho} \approx \frac{\partial V}{\partial \rho}, \quad \frac{\partial P}{\partial \theta} \approx \frac{\partial V}{\partial \theta}.$$

In this case, we arrive at the equations describing the free cylindrical one-dimensional non-relativistic oscillatory movements of electrons in a cold ideal plasma:

$$\frac{\partial V}{\partial \theta} + E + V\frac{\partial V}{\partial \rho} = 0, \quad \frac{\partial E}{\partial \theta} - V + \frac{V}{\rho}\frac{\partial}{\partial \rho}(\rho E) = 0. \tag{4.8}$$

Let's designate it as C1NE (Cylindrical 1-dimension Nonrelativistic Electron oscillations).

Similarly to the relativistic case, we assume that it is required in the semi-infinite strip $\{(\rho, \theta): 0 < \rho < d, \theta > 0\}$ to find the solution of equations (4.8) that satisfies the local initial

$$V(\rho,0) = V_0(\rho), \quad E(\rho,0) = E_0(\rho), \quad \rho \in [0,d], \tag{4.9}$$

and marginal conditions

$$V(0,\theta) = V(d,\theta) = E(0,\theta) = E(d,\theta) = 0 \quad \forall \theta \geqslant 0. \tag{4.10}$$

In this case, we will keep in mind that $V_0(\rho) = 0$, and the function $E_0(\rho)$ is the same in (4.5) and (4.9).

The quasilinear system of equations C1NE (4.8) is important for the construction of numerical algorithms discussed in the chapter, therefore, in addition to writing in Eulerian variables, its form in Lagrangian variables will be useful:

$$\frac{dV(\rho^L,\theta)}{d\theta} = -E(\rho^L,\theta),$$

$$\frac{dE(\rho^L,\theta)}{d\theta} + \frac{E(\rho^L,\theta)V(\rho^L,\theta)}{\rho} = V(\rho^L,\theta),$$

(4.11)

where $d/d\theta = \partial/\partial\theta + V\partial/\partial\rho$ is the total time derivative.

Recall that the function $R(\rho^L, \theta)$, which determines the displacement of a particle with the Lagrangian coordinate ρ^L, so that

$$\rho(\rho^L,\theta) = \rho^L + R(\rho^L,\theta),$$

(4.12)

satisfies the equation

$$\frac{dR(\rho^L,\theta)}{d\theta} = V(\rho^L,\theta).$$

(4.13)

Expressing the speed V through the offset R in accordance with formula (4.13), we write the second equation (4.11) in the form

$$(\rho^L + R)\frac{dE}{d\theta} + E\frac{dR}{d\theta} = (\rho^L + R)\frac{dR}{d\theta}.$$

(4.14)

Equation (4.14) has the first integral

$$(\rho^L + R)E = \frac{1}{2}(\rho^L + R)^2 + C,$$

where the constant C is determined from the condition that the electric field is zero in the absence of particle displacement. Then from this relation we find the expression for the electric field:

$$E(\rho^L,\theta) = \frac{1}{2}\frac{(\rho^L + R(\rho^L,\theta))^2 - (\rho^L)^2}{\rho^L + R(\rho^L,\theta)},$$

(4.15)

and the basic system of equations (4.8) in Lagrangian variables takes the form

$$\frac{dV}{d\theta} = -E, \quad \frac{dR}{d\theta} = V. \tag{4.16}$$

We note that relation (4.12) is very useful for determining the Lagrangian coordinate of the particle ρ^L and the initial condition $R(\rho^L, 0)$ from the given distribution $E_0(\rho)$. The algorithm is as follows: for some ρ from the equation (see the ratio (4.15) with $\theta = 0$)

$$\frac{1}{2} \frac{\rho^2 - (\rho^L)^2}{\rho} = E_0(\rho)$$

by the explicit formula is determined by the value

$$\rho^L = \sqrt{\rho^2 - 2\rho E_0(\rho)}, \tag{4.17}$$

and then from (4.12) the required initial displacement is found at the point ρ^L:

$$R(\rho^L, 0) = \rho - \rho^L. \tag{4.18}$$

To summarize the above, the trajectories of all particles, each of which is identified by the Lagrangian coordinate ρ^L, can be determined by independent integration of the system of ordinary differential equations (4.15), (4.16). For this, two initial conditions are required: $R(\rho^L, 0)$ and $V(\rho^L, 0)$. From (4.7), (4.9) we have $V(\rho^L, 0) = 0$. To determine $R(\rho^L, 0)$, first set the position of the particle ρ at the initial time, then the Lagrangian coordinate is determined by the formula (4.17), and the initial displacement calculated by the formula (4.18). Knowledge of the Lagrangian coordinate ρ^L and the function of the displacement $R(\rho^L, \theta)$ uniquely characterizes the trajectory of the particle by the formula (4.12).

In conclusion of the section, it should be noted that for the relativistic system of equations C1RE (4.4), its analogue in Lagrangian variables has the form

$$\frac{dP(\rho^L, \theta)}{d\theta} = -E(\rho^L, \theta),$$

$$\frac{dR(\rho^L, \theta)}{d\theta} = \frac{P(\rho^L, \theta)}{\sqrt{1 + P^2(\rho^L, \theta)}} \equiv V(\rho^L, \theta), \tag{4.19}$$

the expression for the electric field (4.15) and, therefore, the method of specifying the initial conditions ρ^L, $R(\rho^L, 0)$ and $P(\rho^L, 0) = V(\rho^L, 0) = = 0$ remain unchanged.

4.2. Analytical studies

4.2.1. Axial solution

In [100] (see also Section 2.2), for nonlinear problems describing laser-plasma interactions and possessing axial symmetry, the concept of an axial solution was introduced as a solution having a local-linear dependence on the spatial coordinate.

In this chapter, the axial solution is used to analyze the quality of the considered vibration simulation algorithms. We first establish its useful properties, acting by analogy with [102]. The axial solution of the equations C1NE (4.8) is understood as a real solution of the form

$$V(\rho,\theta) = W(\theta)\rho, \quad E(\rho,\theta) = D(\theta)\rho.$$

It is easy to make sure that the time-dependent factors in this case satisfy the system of ordinary differential equations.

$$W' + D + W^2 = 0, \quad D' - W + 2WD = 0. \tag{4.20}$$

We supplement the equations obtained with arbitrary real initial conditions.

$$W(0) = \beta, \quad D(0) = \alpha \tag{4.21}$$

and find out the conditions for the existence and uniqueness of the solution of the Cauchy problem (4.20), (4.21).

Note that the reduced Cauchy problem is not trivial, since it admits both regular periodic solutions (for example, for small α and β) and solutions that have singularities on a finite time interval (the so-called blow-up solutions). Moreover, the available results even for polynomial right-hand sides (see [137]) in the general case do not allow us to establish the exact boundary between the sets of initial data generating solutions of various types. Therefore, it seems useful to have a slightly different look at the considered statement. Takes place

Lemma 4.2.1. *The Cauchy problem (4.20), (4.21) is equivalent to the following differential algebraic problem:*

$$W' + (\alpha + \beta^2)x + (\alpha - 1/2)x \ln x = 0, \tag{4.22}$$

$$2W^2 + 1 - (1 + 2\beta^2)x - 2(\alpha - 1/2)x \ln x = 0, \tag{4.23}$$

$$W(0) = \beta, \quad x(0) = 1. \tag{4.24}$$

Proof. After excluding the function D from system (4.20), we obtain the Cauchy problem for the second-order equation:

$$W'' + 4W'W + W + 2W^3 = 0,$$
$$W(0) = \beta, \quad W'(0) = -(\alpha + \beta^2). \tag{4.25}$$

Reduce the order of the equation by replacing $p(W) = W'_\theta$:

$$p'_W p + 4pW + W + 2W^3 = 0. \tag{4.26}$$

Hereinafter, the subscript of the derivative clearly indicates the independent variable by which differentiation is carried out. Note that equation (4.26) corresponds (see problem (4.25)) to the initial condition

$$p(\beta) = -(\alpha + \beta^2). \tag{4.27}$$

The transformation of the dependent variable $p(W) = u^{-1}(W) \neq 0$ leads to the equation

$$u'_W - 4u^2W - u^3(W + 2W^3) = 0,$$

in which it is convenient to make the substitution $u(W) = \eta(\xi)$, where $\xi = 2W^2 + C_\xi$. As a result, we will have

$$\eta'_\xi = g(\xi)\eta^3 + \eta^2, \quad g(\xi) = \frac{1}{4}\xi + \frac{1}{4}(1 - C_\xi).$$

To obtain an analytical solution of this equation, we introduce the parameterization of the independent variable $\xi = \xi(t)$ so that

$$\xi'_t = -\frac{1}{t\eta(\xi)}, \quad t \neq 0,$$

and as its corollary we come to the equation

$$t^2 \xi''_t + \frac{1}{4}\xi + \frac{1}{4}(1 - C_\xi) = 0.$$

Its overall solution is

$$\xi(t) = C_1 t^{1/2} + C_2 t^{1/2} \ln t - (1 - C_\xi),$$

where does the formula come from

$$\eta(\xi) = -\left(\frac{1}{2}C_1 t^{1/2} + \frac{1}{2}C_2 t^{1/2} \ln t + C_2 t^{1/2}\right)^{-1}.$$

Returning to the original variables gives

$$p(W) = -\left(\frac{1}{2}C_1 t^{1/2} + \frac{1}{2}C_2 t^{1/2} \ln t + C_2 t^{1/2}\right),$$

$$2W^2 + 1 = C_1 t^{1/2} + C_2 t^{1/2} \ln t.$$

The transformations leading to these relations are completely analogous to those described in detail in the proof of Lemma 2.2.1.

The definition of the constants C_1, C_2 from condition (4.27), that is, from the agreement of the values of the parameters $\theta = 0$ and $t = 1$, and the formal replacement of $t^{1/2} = x$ lead to the differential-algebraic problem (4.22)–(4.24). The lemma is proved.

Note that the procedure performed is an uncomplicated sequence of well-known techniques (see, for example, [62]). Its practical use in this case consists in obtaining an equivalent problem not only in a closed but also in a form more convenient for research. It should be clarified that in this case the equivalence of the considered statements is understood as follows: the function $W(\theta)$ in both formulations is the same, and the remaining functions $D(\theta)$ and $x(\theta)$ are determined in each formulation by $W(\theta)$ uniquely. With the help of a proven statement is established

Theorem 4.2.1. *A necessary and sufficient condition for the existence and uniqueness of a smooth periodic solution of the Cauchy problem* (4.20), (4.21) *is the fulfillment of the inequality*

$$\alpha < 1/2. \qquad (4.28)$$

Proof. *Sufficiency.* Let inequality (4.28) be satisfied. Consider the manifold (4.23). Using linear variable changes

$$z = W\sqrt{2}, \quad x = y\exp(r/s),$$

where $r = 1 + 2\beta^2$, $s = 1 - 2\alpha > 0$, we bring it to the form

$$z^2 + 1 + Ay\ln y = 0 \qquad (4.29)$$

with the parameter $A = s\exp(r/s)$. It is easy to establish that the curve (4.29) is smooth and closed (compact) for $A > e$, and for $A = e$ it degenerates to the point $z = 0$, $y = 1/e$ (here and below, e is the base of the natural logarithm). Moreover, the equality $A = e$ corresponds to the case $\beta = \alpha = 0$, which generates only the trivial solution of the Cauchy problem (4.20), (4.21) and therefore does not represent any interest. For the remaining values of β and $\alpha < 1/2$, the strict inequality $A > e$ holds, i.e.

$$(1 - 2\alpha)\exp\frac{1 + \beta^2}{1 - 2\alpha} > e,$$

which is easily verified by contradiction.

Now, the smoothness and compactness of the manifold (4.23), as well as the smoothness of the right-hand side of system (4.20) in normal form implies the existence and uniqueness of a smooth solution that can be infinitely extended in both directions with respect to $\theta = 0$ (see, for example, [13]) and the Poincaré–Bendixson theorem [66] guarantees its periodicity. Sufficiency is proven.

Necessity. Suppose inequality (4.28) is not satisfied, then two cases are possible. Consider them sequentially.

Let $\alpha = 1/2$, then the manifold (4.23) is a parabola $W^2 = 2y$, where $y = [(1 + 2\beta^2)x - 1]/4$, and the corresponding equation (4.22) has the form

$$W' = -\left(2y + \frac{1}{2}\right) \equiv -W^2 - \frac{1}{2}.$$

The solution of this equation with the initial condition $W(0) = \beta$ is easy to write explicitly:

$$W_0(\theta) = \left[\frac{\beta}{\sqrt{2}} - \frac{1}{2}\tan\frac{\theta}{\sqrt{2}}\right] \Big/ \left[\frac{1}{\sqrt{2}} + \beta\tan\frac{\theta}{\sqrt{2}}\right],$$

whence it follows that it monotonously decreases and becomes unbounded for some $0 < \theta_* < \sqrt{2}\pi$ such that either the denominator is zero for $\beta \neq 0$, or $\theta_* = \pi/\sqrt{2}$ for $\beta = 0$.

Let $\alpha > 1/2$, then equation (4.22) is conveniently written as

$$W' = -(\alpha + \beta^2)x - (\alpha - 1/2)x\ln x.$$

We show that with an arbitrary initial condition $W(0) = \beta$, the solution of this equation decreases faster than $W_0(\theta)$. By the comparison theorem, it suffices to establish the validity of the inequality

$$-(\alpha + \beta^2)x - (\alpha - 1/2)x\ln x < -W^2 - 1/2$$

on the manifold under consideration (4.23). Let the reverse inequality hold

$$-(\alpha + \beta^2)x - (\alpha - 1/2)x\ln x \geqslant -W^2 - 1/2. \tag{4.30}$$

Then from (4.23) we express the value $(\alpha - 1/2)x \ln x$ and substitute it into (4.30). After elementary transformations, we obtain $(1/2 - \alpha) x \ \square \ 0$. Now we recall that $x > 0$ (see the proof of the lemma), from which we obtain the required contradiction with the assumption of the value of α, i.e. the validity of the majorant estimate for any values of W. Finally, we see that the failure to satisfy condition (4.28) implies that the solution is unbounded on a finite time interval, that is, the blow-up property. The theorem is proved.

We will comment on the result obtained from two points of view: the physical interpretation of the functions and the finding of an approximate (numerical) solution of the problem (4.20), (4.21).

For the Cauchy problem under study, we consider the simplest unbounded solution $W_0(\theta)$ that satisfies the equation

$$W' = \frac{-(2W^2 + 1)}{2}.$$

This implies the formula $D_0(\theta) \equiv 1/2$, and, accordingly, the expression for electron density $N_0 = 1 - 2D_0 \equiv 0$. Thus, from a physical point of view, the critical value $\alpha = 1/2$ is meaningless, since it means the absence electrons (their concentration is zero). In other words, oscillating objects do not exist.

From the point of view of numerical methods, it should be noted that the differential-algebraic formulation (4.22)–(4.24) has no visible advantages as compared with the fully differential formulation (4.20), (4.21). On the contrary, the Cauchy problem (4.20), (4.21), which has a smooth periodic solution, can be successfully numerically integrated by a variety of methods (see, for example, [16, 20]), but the convergence of an approximate solution to the exact one can be accurately justified only under the condition the stability of the solution with respect to the initial data, which is closely related to the smoothness and compactness of the manifold (4.23).

4.2.2. Perturbation method

If we assume that the oscillation amplitude is sufficiently small, that is, in the initial condition (4.7) for the electric field $a_* \; \Box \; \rho_*$, then the equations C1NE (4.8) become weakly nonlinear and their approximate solutions can be constructed using the perturbation theory technique [25, 154] (see also p. 3.2.1). We give here a brief derivation of the corresponding analytical formulas.

Eliminating the electric field E from system (4.8), we obtain the following equation for the velocity V_p:

$$\left(\frac{\partial^2}{\partial \theta^2} + 1\right)V_p + \frac{\partial}{\partial \theta}\left(V_p \frac{\partial V_p}{\partial \rho}\right) + V_p \frac{1}{\rho}\frac{\partial}{\partial \rho}\left(\rho \frac{\partial V_p}{\partial \theta}\right) +$$

$$+ V_p \frac{1}{\rho}\frac{\partial}{\partial \rho}\left(\rho V_p \frac{\partial V_p}{\partial \rho}\right) = 0. \tag{4.31}$$

Hereinafter, the subscript p will indicate approximations sought in accordance with perturbation theory.

Equation (4.31) will be solved, considering the nonlinear terms to be small. Substituting the velocity of electrons in the form of expansion in powers of nonlinearity

$$V_p = V_1 + V_2 + V_3 + \ldots,$$

we find that the first approximation satisfying the initial conditions

$$V_1\big|_{\theta=0} = 0, \quad \frac{\partial V_1}{\partial \theta}\bigg|_{\theta=0} = A(\rho),$$

has the appearance

$$V_1 = A\sin\theta, \tag{4.32}$$

where $A = A(\rho)$ is the amplitude of the electron velocity oscillations depending on the radius.

For the second and third terms of the asymptotic expansion of the solution (4.31), we successively obtain the equations

$$\left(\frac{\partial^2}{\partial \theta^2} + 1\right)V_2 + \frac{\partial}{\partial \theta}\left(V_1 \frac{\partial V_1}{\partial \rho}\right) + V_1 \frac{1}{\rho}\frac{\partial}{\partial \rho}\left(\rho \frac{\partial V_1}{\partial \theta}\right) = 0, \tag{4.33}$$

$$\left(\frac{\partial^2}{\partial \theta^2} + 1\right)V_3 - 2\omega_2 V_1 + \frac{\partial}{\partial \theta}\left(V_1 \frac{\partial V_2}{\partial \rho} + V_2 \frac{\partial V_1}{\partial \rho}\right) + V_1 \frac{1}{\rho}\frac{\partial}{\partial \rho}\left(\rho \frac{\partial V_2}{\partial \theta}\right) +$$
$$+ V_2 \frac{1}{\rho}\frac{\partial}{\partial \rho}\left(\rho \frac{\partial V_1}{\partial \theta}\right) + V_1 \frac{1}{\rho}\frac{\partial}{\partial \rho}\left(\rho V_1 \frac{\partial V_1}{\partial \rho}\right) = 0. \tag{4.34}$$

Note that the particular solution of equation (4.33) with (4.32) taken into account is

$$V_2 = \frac{1}{2}\left(A\frac{dA}{d\rho} + \frac{A^2}{3\rho}\right)\sin 2\theta. \tag{4.35}$$

Now, to eliminate the resonant terms in equation (4.34), which lead to time-escalating solutions, one should take into account the frequency change in formula (4.32):

$$V_1 = A\sin\omega\theta, \quad \omega = 1 + \omega_2, \tag{4.36}$$

moreover, ω_2 is a quadratic in amplitude correction to the fundamental frequency. We use the explicit formulas for V_1, V_2 and equate to zero the coefficient for $\sin \omega\theta$ in (4.34). As a result, we get

$$\omega_2 = \frac{A^2}{12\rho^2}. \tag{4.37}$$

This expression is not new, as it was already encountered in [131]. In this case, of greater interest is the formula for electron density, which follows from expression (4.36) and the approximation $V_p \approx V_1$:

$$N_p = 1 + \frac{1}{\rho}\frac{d}{d\rho}(\rho A)\cos\omega\theta - \theta\frac{d\omega}{d\rho}A\sin\omega\theta.$$

Such a representation means that from the asymptotic representation of the velocity, it is sufficient to hold only the first term, although the nonlinear frequency shift (4.37) follows from the regular behavior of the three terms in the expansion in a small parameter of the oscillation amplitude.

4.3. Finite difference method

The system of equations C1NE (4.8), in essence, represents the interaction of two physical processes: nonlinear oscillations at a fixed point in space and their space-time transfer. Therefore, it is convenient to preface the formal description of the proposed schemes with a reminder of basic auxiliary structures. In this case, the central place is given to the time discretization procedure in order to obtain a second order approximation.

4.3.1. Auxiliary designs
Splitting into physical processes. Consider the model equation

$$\frac{\partial u}{\partial t} = L_1(u) + L_2(u), \tag{4.38}$$

where L_i, $i = 1, 2$, is generally nonlinear operators, possibly (but not necessarily) associated with differentiation over space. We introduce time discretization: $t_j = j\tau$, $j \square 0$, τ is the step in the variable t. For simplicity, we assume that the solution depends only on two independent variables t and x. It is convenient to use the following notation: $u = u(t_j, x)$, $\hat{u} = u(t_j + \tau, x)$. Then, assuming smoothness of the solution (4.38), we will have

$$\hat{u} = u + \tau[L_1(u) + L_2(u)] + O(\tau^2).$$

Now, in accordance with [41, 74], we replace the solution (4.38) on the interval $[t_j, t_j + \tau]$ with the sequential solution of two auxiliary problems:

$$\frac{\partial v}{\partial t} = L_1(v), \quad v|_{t=t_j} = u \tag{4.39}$$

and

$$\frac{\partial w}{\partial t} = L_2(w), \quad w|_{t=t_j} = \hat{v}. \tag{4.40}$$

Note that in the case of sufficient smoothness of the desired solution and the polynomial (in our case quadratic) nonlinearity of the operators L_i, $i = 1,2$, we have $\hat{w} = \hat{u} + O(\tau^2)$.
Really,

$$\hat{v} = v + \tau L_1(v) + O(\tau^2) = u + \tau L_1(u) + O(\tau^2),$$
$$\hat{w} = w + \tau L_2(w) + O(\tau^2) = \hat{v} + \tau L_2(\hat{v}) + O(\tau^2) =$$
$$= u + \tau L_1(u) + \tau L_2(u + \tau L_1(u)) + O(\tau^2) =$$
$$= u + \tau[L_1(u) + L_2(u)] + O(\tau^2) = \hat{u} + O(\tau^2).$$

For the actual solution of equations (4.39) and (4.40), we will further use the approximations of order $O(\tau^2 + h^2)$, where h is the discretization step in the spatial variable.

Lax–Wendroff scheme ('tripod'). Consider another model equation (such as non-linear transfer)

$$\frac{\partial u}{\partial t} + \frac{\partial F(u)}{\partial x} = 0, \quad F(u) = \frac{u^2}{2},$$

and introduce the notation $A = \dfrac{\partial F}{\partial u}$. Assuming that the solution u is sufficiently smooth, we have

$$u_{tt} = -F_{xt} = -F_{tx}, \quad F_t = F_u u_t = -F_u F_x \equiv -AF_x.$$

This gives $u_{tt} = (AF_x)_x$ and, respectively,

$$\hat{u} = u + \tau u_t + \frac{\tau^2}{2} u_{tt} + O(\tau^3) = u - \tau F_x + \frac{\tau^2}{2}(AF_x)_x + O(\tau^3).$$

In the above calculations, the subscript denotes the partial differentiation with respect to the corresponding variable. Now you can record the time sampling of the 'tripod' scheme

$$\frac{u^{j+1} - u^j}{\tau} + \frac{\partial F^j}{\partial x} = \frac{\tau}{2} \frac{\partial}{\partial x}\left(A^j \frac{\partial F^j}{\partial x} \right),$$

having an $O(\tau^2)$ approximation. Here the superscript denotes the belonging of the function to the corresponding moment of time.

Note that if we take the linear transport equation as a model equation

$$\frac{\partial u}{\partial t} + v \frac{\partial u}{\partial x} = 0,$$

then for him time discretization, as in the «tripod» scheme, will have the form

$$\frac{u^{j+1} - u^j}{\tau} + \left(v^j + \frac{\tau}{2} \frac{\partial v}{\partial t} \right) \frac{\partial u}{\partial x} = \frac{\tau v^j}{2} \frac{\partial}{\partial x}\left(v^j \frac{\partial u}{\partial x} \right),$$

and also have an $O(\tau^2)$ approximation on smooth solutions.

4.3.2. Construction of difference schemes

We present equations (4.8) to a convenient form. To this end, we single out the explicit term in the second equation, which is responsible for the shift of the oscillation frequency. As a result, we get

$$\frac{\partial V}{\partial \theta} + E + V\frac{\partial V}{\partial \rho} = 0, \quad \frac{\partial E}{\partial \theta} - V + \frac{VE}{\rho} + V\frac{\partial E}{\partial \rho} = 0.$$

We now relate to the description of the process of nonlinear oscillations of the equation

$$\frac{\partial \tilde{V}}{\partial \theta} + \tilde{E} = 0, \quad \frac{\partial \tilde{E}}{\partial \theta} - \tilde{V} + \frac{\tilde{V}\tilde{E}}{\rho} = 0, \tag{4.41}$$

and to their transfer in space and time –

$$\frac{\partial \overline{V}}{\partial \theta} + \overline{V}\frac{\partial \overline{V}}{\partial \rho} = 0, \quad \frac{\partial \overline{E}}{\partial \theta} + \overline{V}\frac{\partial \overline{E}}{\partial \rho} = 0. \tag{4.42}$$

As a basis for time sampling of both systems apply the usual leapfrog scheme (see Section 3.5). Let τ – time step, then we will refer to the 'whole' moments of time $\theta_j = j\tau$ ($j \geq 0$ is an integer) of E, $\tilde{E}$, $\overline{E}$, N, and to the 'half-integer' $\theta_{j\pm 1/2}$ – V, $\tilde{V}$, $\overline{V}$. The choice of the corresponding time moment for the function value will be denoted by superscript. For discretization in space, we will use a grid with a constant step h so that $\rho_m = mh$, $0 \leq m \leq M$, $Mh = d$.

We write the difference equations approximating systems (4.41) and (4.42). For the first one we get

$$\frac{\tilde{V}_m^{j+1/2} - \tilde{V}_m^{j-1/2}}{\tau} + \tilde{E}_m^j = 0,$$

$$\frac{\tilde{E}_m^{j+1} - \tilde{E}_m^j}{\tau} - \tilde{V}_m^{j+1/2} + \frac{\tilde{V}_m^{j+1/2}}{\rho_m}\frac{\tilde{E}_m^{j+1} + \tilde{E}_m^j}{2} = 0, \tag{4.43}$$

$$\tilde{V}_m^{j-1/2} = V_m^{j-1/2}, \quad \tilde{E}_m^j = E_m^j, \quad 1 \leqslant m \leqslant M - 1.$$

The discrete analogue of the system (4.42) has the following form:

$$\frac{\overline{V}_m^{j+1/2} - \overline{V}_m^{j-1/2}}{\tau} + F_{\check{X},m}^{j-1/2} = \frac{\tau}{2}\left(\overline{V}_{s,m}^{j-1/2} F_{x,m}^{j-1/2}\right)_{\overline{X},m},$$

$$\frac{\overline{E}_m^{j+1} - \overline{E}_m^{j}}{\tau} + \left(\overline{V}_m^{j+1/2} + \frac{\tau}{2}\frac{\overline{V}_m^{j+1/2} - \overline{V}_m^{j-1/2}}{\tau}\right)\overline{E}_{\check{X},m}^{j} =$$

$$= \frac{\tau}{2}\overline{V}_m^{j+1/2}\left(\overline{V}_{s,m}^{j+1/2} \overline{E}_{x,m}^{j}\right)_{\overline{X},m},$$

$$\overline{V}_m^{j-1/2} = \tilde{V}_m^{j+1/2}, \quad \overline{E}_m^{j} = \tilde{E}_m^{j+1}, \quad 1 \leqslant m \leqslant M-1,$$

$$\overline{V}_0^{j+1/2} = \overline{V}_M^{j+1/2} = \overline{E}_0^{j+1} = \overline{E}_M^{j+1} = 0.$$

(4.44)

In expression (4.44), the following notation is used: $F^{j-1/2} \equiv F(\overline{V}^{j-1/2}) = \frac{1}{2}\left(\overline{V}_m^{j-1/2}\right)^2$, $F_{\check{X},m} = (F_{m+1} - F_{m-1})/(2h)$ − central difference, $F_{X,m} = (F_{m+1}-F_m)/h$ and $F_{\overline{X},m} = (F_m - F_{m-1})/h$ − differences forward and backward, respectively, $F_{s,m} = (F_{m+1} + F_m)/2$.

After calculations using the scheme (4.44), you should redefine the required functions at the following time layer:

$$V_m^{j+1/2} = \overline{V}_m^{j+1/2}, \quad E_m^{j+1} = \overline{E}_m^{j+1}, \quad 0 \leqslant m \leqslant M,$$

and calculate (if necessary) the value of the electron density by the formula

$$N_m^{j+1} = \begin{cases} 1 \text{ де } \dfrac{1}{\rho_m}\dfrac{\rho_{m+1}E_{m+1}^{j+1} - \rho_{m-1}E_{m-1}^{j+1}}{2h}, & \leqslant m \leqslant M- \\[2ex] 1 \text{ д } 2\dfrac{E_1^{j+1}}{h} & \text{è } \quad m = 0, \\[2ex] 1 \text{ д} & \text{è } \quad m = M. \end{cases}$$

(4.45)

At this time, the calculations at the jth time step end and you can proceed to the next step. It should be noted that the initial data (4.9) correspond to $j = 0$, therefore they should be attributed for V to the layer with the number $-1/2$, and for E to the layer with the number 0.

We make a remark about the splitting scheme under consideration (4.43), (4.44). For each of the auxiliary problems with sufficient smoothness of the solution, there is an approximation of order

$O(\tau^2 + h^2)$, as well as the stability condition obtained on the basis of the spectral attribute [7, 20], of the form $\tau = O(h)$. This is already a good argument for the fact that the scheme under consideration is more convenient for calculations than those previously used for the same purposes [45], [123]. In particular, the use of a new scheme makes it possible to achieve significant savings in computational resources due to a weaker stability condition without loss of approximation. In addition, the scheme (4.43), (4.44) is explicit, which generates the potential for parallelization when generalized to multidimensional cases.

4.3.3. Process scenario

The formal formulation of the problem in the Eulerian variables (4.3), (4.8), (4.9), (4.10), allows us to generally characterize the qualitative scenario of the dynamics of axially symmetric plasma oscillations. For definiteness, we fix the values of the parameters in (4.7): $a_* = 0.365$, $\rho_* = 0.6$, and consider the corresponding fig. 4.1. It clearly shows the process of evolution of nonlinear plasma oscillations in time, as well as their radial structure. At the initial stage of oscillation, the maximum electron density is located on the axis and exceeds the background value by about an order of magnitude. Further, over time, in the process of oscillation, two trends are observed. The first of these is that off-axis oscillations are somewhat lagging in phase from oscillations on the axis and from phase to period this phase shift increases. The second trend is more obvious: over time, there is a gradual formation of the absolute maximum density, located off-axis and comparable in magnitude with the axial ones.

The described picture with two maxima is shown separately in Fig. 4.2, with the viewing angle being changed for clarity. The formation of the off-axis maximum of the electron density is a signal that the regular development of oscillations has ended and the process of their destruction has begun. The dynamics of destruction is most noticeably manifested in the growth of the value of the off-axis extremum: from period to period it grows and eventually turns into infinity, i.e. oscillation breaks. With these parameters, the off-axis density maximum was formed at $\theta_{\max}^{(1)} \approx 34$, increased approximately twice in the next period at $\theta_{\max}^{(2)} \approx 40$, and turned into infinity at $\theta_{\mathrm{br}} \approx 46$. By virtue of the validity of formula (4.3), the singularity density means the formation of a jump (discontinuity) of the radial

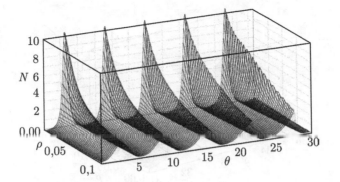

Fig. 4.1. Spatio-temporal distribution of electron density before the formation of an off-axis maximum

component of the electric field. Note that the axial extremes of the electron density in this case behave in a regular manner, i.e., their absolute values and the periodicity of appearance are almost unchanged up to breaking.

We now fix some value $\rho_* < 1$ and characterize the variations of the described process as the parameter a_* changes. Suppose first that a_* decreases monotonically. Then the evolution of plasma oscillations will stretch in time, approaching asymptotically to the results of a weakly nonlinear model, which was studied in detail in [46, 123] (see also [45]). Recall that for $\rho_* \square 1$ and sufficiently small a_*, asymptotic formulas are valid for the amplitude of the oscillations $N_{max} = C_1(a_*/\rho_*)^2$, the formation time of the first off-axis maximum of the electron density $\theta_{max}^{(1)} = C_2(\rho_* / a_*)^4$ and oscillation breaking time $\theta_{br} = C_3(\rho_*/a_*)^6$ with some constants C_i, $i = 1, 2, 3$. For this approximation, the radial coordinate of the rollover point is also known – $\rho_{br} = \rho_* / \sqrt{6}$.

Let the parameter a_* now increase monotonically. Then, on the contrary, the evolution of plasma oscillations will shrink with time, acquiring an increasingly non-linear character. First of all, this will be noticeable in absolute values of regular axial maxima: their values will begin to exceed the background by a factor of ten or more. For example, for $a_* = 0.391$, we have $N_{axis} \approx 111$, and for $a_* = 0.401$, we have $N_{axis} \approx 1244$. The oscillation breaking time and the corresponding radial coordinate will decrease. At the same time there is such a high-quality picture. If breaking oscillations occur within a single period, then their radial coordinate decreases monotonically with increasing a_*. When «skipping» over the time of rollover in the previous period, the radial coordinate increases abruptly and then gradually decreases inside the period. At the same time, its minimum

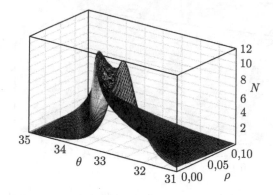

Fig. 4.2. Spatio-temporal distribution of electron density during the formation of an off-axis maximum

value for periods tends to zero, that is, to the axis. During the growth of the parameter a_* (for a fixed value of ρ_*), it is not always possible to trace the formation and growth of the off-axis density extremum. Off-axis breaking often occurs so fast that the off-axis extremum cannot survive even one period in time. We also note that the formation time of the first axial extremum of electron density also decreases with increasing a_*. At least, never in the calculations was not detected breaking effect on the eve of the appearance of the first axial maximum. The critical value in the hydrodynamic model is $(a_*/\rho_*)^2 = 1/2$, in the vicinity of which the breaking effect will be almost axial: the initial distribution of electrons is such that they all rush towards the axis, are reflected from it, and then very quickly an intersection of electron trajectories occurs. Thus, the duration of oscillations in the vicinity of the critical value will not exceed one period. This fact follows from the study of axial solutions of the problem under consideration [100] (see also section 4.2.1).

Note that there is another scenario of the dynamics of electronic oscillations, significantly different from the one discussed above. It takes place with a negative distribution of the electric field at the initial moment of time (i.e., if we change the sign in the formula (4.7)). For example, for $a_* = 0.65$, $\rho_* = 0.6$, we have a completely different picture of the electron density. On an axis with a period of approximately 2π, small identical maxima $N_{axis} \approx 3.3$ are located (the first is immediately at $\theta = 0$), and outside the axis at a distance of approximately $\rho = 0.78$ with a time shift of about half the period, a sequence of monotonically increasing to infinity) off-axis maxima (see fig. 4.3). Of course, their regularity in time and space

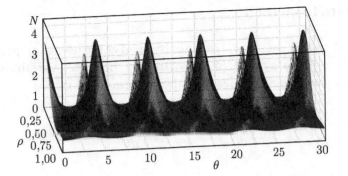

Fig. 4.3. Spatio-temporal distribution of electron density with oppositely directed electric field

is different from axial ones. They gradually approach each other and shift towards the axis, but these movements are rather smooth and do not lead to singularities. At the same time, the first value of the off-axis maximum is about one and a half times the axial, but breaking occurs much later than in the first scenario: $\theta_{br} \approx 114$. There is another important difference: the radial coordinate of breaking is far enough from the axis (in this case $- \rho_{br} \approx 0.36$), and this sharply reduces the computational difficulty of the problem. However, we note that the main goal of the chapter is to study the process of oscillations, which is similar to the dynamics of wake waves excited by a laser pulse, so the first scenario is of priority importance, which this chapter is devoted to.

From the description of the oscillation scenario, it is easy to understand why it is almost impossible to model the breaking process in Eulerian variables without using specially designed schemes. Successful calculations are equally hindered by both of these tendencies: both the curvature of the phase front and the unlimited growth of the off-axis maximum. As a result of their development, the smoothness of the desired functions deteriorates to such an extent that it requires substantial refinement of the discretization parameters in time and space simultaneously. The latter leads to the claim of colossal computational resources, but even their use does not lead to success, since the indicated tendencies continue to develop up to the breaking effect. This implies why the numerical solution of problem (4.3), (4.8), (4.9), (4.10) using traditional difference schemes [45, 123] did not allow studying the structure of the electron density after the appearance of an off-axis maximum in the mode of even moderate nonlinearity.

4.4. Particle method

From the point of view of the hydrodynamic description, an alternative to the system (4.8) written in Eulerian variables is the use of Lagrangian variables, i.e. the system (4.11):

$$\frac{d}{d\theta}V=-E, \quad \frac{d}{d\theta}E+V\frac{E}{\rho}=V, \quad \frac{d}{d\theta}R=V, \tag{4.46}$$

where $\dfrac{d}{d\theta}=\dfrac{\partial}{\partial\theta}+V\dfrac{\partial}{\partial\rho}$ is the total time derivative, R is the displacement that determines the trajectory of the particles

$$\rho(\rho^L,\theta)=\rho^L+R(\rho^L,\theta),$$

whose initial position in the radial coordinate is ρ^L.

From the system (4.11) in section. 4.1 was obtained explicitly the relationship between the electric field and the displacement

$$E=\frac{1}{2}\frac{(\rho^L+R)^2-(\rho^L)^2}{\rho^L+R}. \tag{4.47}$$

Using this expression allows us to obtain simpler equations for determining the trajectories of particles:

$$\frac{dR}{d\theta}=V, \quad \frac{dV}{d\theta}=-E, \tag{4.48}$$

as well as the initial conditions for their solution. Let us pay attention to the fundamental difference between the systems (4.8) and (4.48): in the latter, there is no explicit partial derivative with respect to the radial coordinate, i.e., system (4.48) is actually formally independent ordinary differential equations for individual particles. Of course, in fact, there is a connection between the initial conditions for these equations. In addition, for a visual representation of the process, numerical differentiation by the radial variable of the electric field function (4.47) is required in accordance with formula (4.3). However, the situation has changed qualitatively: the desired displacement functions of particles $R(\rho^L, \theta)$ are smooth functions of the variable θ, and depend on ρ^L as an external (initial) parameter. This circumstance makes the numerical solution of system (4.48) a very easy procedure.

For completeness of the description, we present the formulas of the computational algorithm. Suppose that in the initial period of time $\theta = 0$, the particle with the number k is characterized by the initial position along the radius $\rho_0(k)$ and the initial displacement $R(k, 0)$, where $1 \le k \le M$, M is the total number of particles. The initial positions of all particles form the electric field, which has the form (4.7). On the other hand, the deviation of particles at the initial moment of time creates at the point with the coordinate $\rho_k = \rho_0(k) + R(k, 0)$ an electric field in accordance with formula (4.47). Comparing expressions (4.7) and (4.47), we can determine the desired values of $\rho_0(k)$ and $R(k, 0)$. To do this, we define the initial spatial grid $\rho_k = kh$, h is the discretization parameter with respect to the radial variable, which characterizes the proximity of neighboring particles. In grid nodes, using formula (4.7), we calculate the values of the electric field $E(\rho_k, 0)$. This electric field is formed by the displacement of particles, that is, on the basis of (4.47) we have the equations for determining the initial positions $\rho_0(k)$:

$$E(\rho_k, 0) = \frac{1}{2} \frac{\rho_k^2 - \rho_0^2(k)}{\rho_k},$$

and then, recalling that $\rho_k = \rho_0(k) + R(k, 0)$, from the initial positions of the particles already found, we find their initial deviations $R(k, 0)$ from these initial positions. Thus, to calculate the trajectory of each particle, initial data are obtained, to which we must add the condition of particle immobility at the initial moment of time from (4.7), that is, $V(k, 0) = 0$.

Note that in the vicinity of the axis, the dependence of the electric field E on the radius is close to linear, i.e. $E \approx \alpha\rho$, $\alpha = (a_*/\rho_*)^2$. Therefore, when specifying the initial positions of the particles, the formula $\rho_0(k) \approx \rho_k \sqrt{1 - 2\alpha}$ is valid, which, in turn, implies the criticality of $\alpha = 1/2$ in order to preserve the correctness of the hydrodynamic description of the process.

Equations (4.48), as noted above, are ordinary differential equations. Therefore, they can be integrated in a numerically usual way [20], for example, according to the traditional (see [139]) for equations of motion, a second-order scheme of accuracy (the so-called leapfrog scheme). Let τ be a time discretization parameter, i.e. $\theta_j = j\tau$, $j \geq 0$, then the calculation formulas will have the following form:

$$\frac{V(k,\theta_{j+1/2}) - V(k,\theta_{j-1/2})}{\tau} = -\frac{1}{2}\frac{(\rho_0(k) + R(k,\theta_j))^2 - \rho_0^2(k)}{\rho_0(k) + R(k,\theta_j)},$$

$$\frac{R(k,\theta_{j+1}) - R(k,\theta_j)}{\tau} = V(k,\theta_{j+1/2}).$$

In this case, at an arbitrary time θ_j, the variable Eulerian grid can be calculated from the formula

$$\rho_k = \rho_0(k) + R(k,\theta_j), \quad 1 \leq k \leq M,$$

in the nodes of which, in accordance with (4.47), the values of the electric field $E(\rho k, \theta_j)$ are determined. This is used to represent electron density for illustrative purposes: the calculations use the formula for the numerical differentiation of the second order of accuracy in the middle of the sub-segments

$$N\left(\frac{\rho_{k+1} + \rho_k}{2}, \theta_j\right) =$$

$$= 1 - \left[\frac{E(\rho_{k+1},\theta_j) - E(\rho_k,\theta_j)}{\rho_{k+1} - \rho_k} + \frac{E(\rho_{k+1},\theta_j) + E(\rho_k,\theta_j)}{\rho_{k+1} + \rho_k}\right].$$

Here it is convenient to assume that at any moment of time on the axis $\rho = 0$ there is a particle with the number $k = 0$, which always has no displacement, that is, its trajectory simply coincides with the axis and on this trajectory the electric field is always zero.

Calculations according to the presented scheme were carried out for the number of particles $M \leq 5000$ with the same parameters as when using the Eulerian variables:

$$a_* = 0.365, \quad \rho_* = 0.6, \quad h = 1/1600, \quad \tau = 1/32000.$$

The calculations performed using Lagrangian variables completely reproduce the previously obtained results for axial density maxima.

In addition, there is good agreement for the magnitude and position of the first off-axis electron density maximum. In fig. 4.4 presents the density distribution at the moment of the formation of the first off-axis maximum at $\theta_{max}^{(1)} \approx 34$. This maximum is at a distance from the axis equal to $\rho_{max}^{(1)} \approx 0.046$, and its value is $N_{max}^{(1)} \approx 9.5$. The spatial structure of the electric field and the electron velocity are also calculated. In fig. 4.5 presents the results of these calculations at the time of the appearance of the first off-axis maximum density. The approach described above, based on Lagrangian variables, allows one to move further in time and investigate the structure of the electron density after the appearance of the first off-axis maximum. Numerical calculations show that over time, the density maxima located on the axis do not change their magnitude. In contrast, the off-axis maximum of the density after its appearance at the moment of time $\theta_{max}^{(1)}$ quickly increases from period to period. The second off-axis density maximum has the value $N_{max}^{(2)} \approx 23$ and arises at the time instant $\theta_{max}^{(2)} \approx 40$ at a distance from the axis equal to $\rho_{max}^{(2)} \approx 0.053$. And finally, in the next period of oscillations, at the instant of time $\theta_{max}^{(3)} \approx 46$, at a distance from the axis $\rho_{max}^{(3)} \approx 0.035$, the density becomes infinite, that is, cylindrical plasma oscillations overturn. It is with this sharp increase in the off-axis maximum of the density that the difficulties of numerical integration of the system of partial differential equations (4.8) – the system with the use of Eulerian variables – are associated. Calculations also show that with the passage of time, the radial profile of the electric field becomes steeper and a fracture forms at the electron velocity. When plasma oscillations break at the instant of time $\theta_{max}^{(3)}$, a jump in the electric field is observed, and the velocity of the electrons has a discontinuity in the radial derivative (see Fig. 4.6). The calculations of the particle trajectories for the above parameters show that in the process of oscillatory motion, the distances between the neighbouring particles vary depending on their initial positions. The trajectories of some neighbouring particles converge, and at a certain point in time they intersect. In this case, the instants of time for the intersection of the particle trajectories and the density to infinity exactly coincide.

Note that, in contrast to the use of Eulerian variables, where the computational domain was taken to be rather large, $0 \square \rho \square d \approx 4.5\rho_*$, in order to satisfy the boundary conditions (4.10), there is no need for Lagrangian variables. In this case, it suffices to take into account only particles that will later be in the breaking zone. The

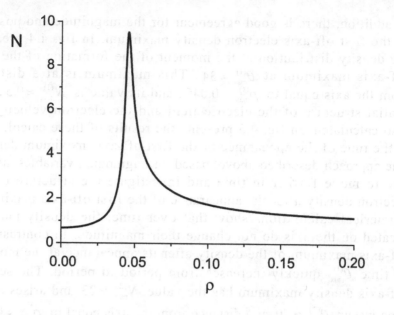

Fig. 4.4. Spatial distribution of electron density at the moment of formation of the first off-axis maximum.

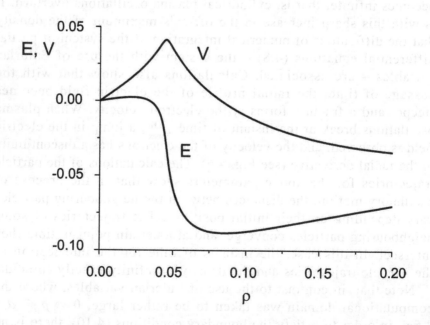

Fig. 4.5. Spatial distribution of velocity and electric field at the moment of formation of the first off-axis maximum.

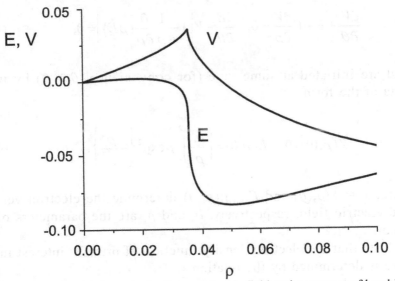

Fig. 4.6. Spatial distribution of velocity and electric field at the moment of breaking of plasma oscillations

area of space occupied by them is rather small — less than 10% of ρ_* and less than 2.5% of d. Taking into account this circumstance allows to significantly increase the efficiency of the computational algorithm in determining the coordinates of plasma oscillations breaking due to a simultaneous decrease in the amount of calculations (reduction in the number of particles) and an increase in the accuracy of calculations (decrease in the h parameter).

4.5. Calculation of axial solutions

The section deals with finding axial solutions for two problems associated with cylindrical plasma oscillations. For each of them, the formulation of the 'complete' problem is given, then equations for axial solutions are derived, then the numerical method of their integration is described. In addition, the axial solutions obtained on the basis of calculations are compared with solutions on the axis of 'complete' problems that were obtained by other methods.

4.5.1. Free non-relativistic oscillations

Recall the formulation of the 'complete' problem. In the nonrelativistic case, free cylindrical oscillations are described by the system of equations C1NE:

$$\frac{\partial V}{\partial \theta}+E+V\frac{\partial V}{\partial \rho}=0, \quad \frac{\partial E}{\partial \theta}-V\left[1-\frac{1}{\rho}\frac{\partial}{\partial \rho}(\rho E)\right]=0, \tag{4.49}$$

and are initiated at some time (for convenience, $\theta = 0$) by initial data of the form

$$V(\rho,0)=0, \quad E(\rho,0)=\left(\frac{a_*}{\rho_*}\right)^2 \rho \exp^2\left\{-\frac{\rho^2}{\rho_*^2}\right\}. \tag{4.50}$$

Here, $V = V(\rho, \theta)$ and $E = E(\rho, \theta)$ determine the electron velocity and electric field, respectively, a_* and ρ_* are the parameters of the problem.

Note that the electron density function of primary interest in this case is determined by the relation

$$N(\rho,\theta)=1-\frac{1}{\rho}\frac{\partial}{\partial \rho}(\rho E). \tag{4.51}$$

To unambiguously determine the desited functions, one needs to add boundary conditions. On the axis (at $\rho = 0$), due to the axial symmetry of the problem, we have

$$V(0,\theta)=E(0,\theta)=0. \tag{4.52}$$

In addition, the physical formulation provides that the forcing electric field (4.50) rapidly decreases in the radial direction. Therefore, for a sufficiently large value of $\rho = d$, such that $\exp(-d^2/\rho_*^2)\ll 1$, the equations are valid with satisfactory accuracy

$$V(d,\theta)=E(d,\theta)=0. \tag{4.53}$$

Thus, the simplest problem of cylindrical plasma oscillations is formulated as follows: find in the field

$$\Omega=\left\{(\rho,\theta):0\leqslant\rho\leqslant d, 0\leqslant\theta\leqslant\theta_{max}\right\}$$

functions V and E, satisfying equations (4.49), initial data (4.50) and boundary conditions (4.52), (4.53). The value of θ_{max} is determined only by the interests of the study. We note the characteristic feature of the statement: of primary interest is the electron density function on the axis (with $\rho = 0$), but for this it is necessary to solve the 'complete' problem.

To calculate only the axial solution, we formulate the 'truncated' problem. Taking into account the form of the boundary conditions (4.52), in the vicinity of the axis we will consider solutions of equations (4.49) that are linear in space, i.e.

$$V(\rho,\theta) = W(\theta)\rho, \quad E(\rho,\theta) = D(\theta)\rho.$$

It is easy to verify that in this case the partial differential equations (4.49) go over to ordinary differential equations of the form

$$W' + D + W^2 = 0, \quad D' - W(1 - 2D) = 0, \tag{4.54}$$

and the initial conditions (4.50) – respectively in the conditions

$$W(0) = 0, \quad D(0) = (a_* / \rho_*)^2. \tag{4.55}$$

In addition, for the electron density function we have an expression on the axis

$$N(\rho = 0, \theta) = 1 - 2D(\theta).$$

Thus, the problem of determining axial solutions in this case reduces to the integration of equations (4.54) in the time interval $0 \le \theta \le \theta_{max}$, starting with the initial conditions (4.55).

We give a description of the numerical scheme and the results of calculations. We introduce a uniform grid $\theta_k = k\tau$, $0 \le k \le K$, $K\tau = \theta_{max}$ and write the discrete analogue of equations (4.54). Using the notation $f_k = f(\theta_k)$ for grid functions, we will have for $k \ge 0$ the scheme

$$\frac{W_{k+1} - W_k}{\tau} + D_k + W_k^2 = 0,$$

$$\frac{D_{k+1} - D_k}{\tau} - W_{k+1}(1 - 2D_{k+1}) = 0. \tag{4.56}$$

Difference initial conditions are determined in a natural way:

$$W_0 = 0, \quad D_0 = (a_* / \rho_*)^2.$$

The initial equations have non-linear terms, but the chosen scheme is non-iterative: both functions are calculated using explicit formulas. It is easy to see that in the nonlinear case for smooth solutions the scheme has the first order of accuracy, and for the linearized formulation the second. Of course, due to the simple structure (4.54), it is possible to propose methods of a higher order [22], however, the simplicity of the calculations and the simple generalization of the scheme to the case of solving the 'complete' problem make formulas (4.56) quite attractive.

Consider the characteristic calculation option according to the scheme (4.56); the images in fig. 4.7 and 4.8 of the electron density N and the derivative of the velocity W in the region $0 \square \theta \square 50.3$ with the following values of the parameters: $a_* = 0.365$, $\rho_* = 0.6$. For comparison, we take the calculation of the "complete" problem (4.49), (4.50) with the indicated parameters from Sec. 4.3 (see also [45]). In order to save space, we will not reproduce the detailed description of the difference scheme used there, but for a formally correct presentation we add the missing value $d = 2.7$.

The comparison results are shown in Table. 4.1, which at specially selected times θ contains extreme (minimum and maximum) electron density values on the axis $\rho = 0$, calculated in various ways. The columns $N_{f,1}$ and $N_{f,2}$ are obtained by the finite difference method from the solution of the 'complete' problem for the following values of the grid parameters:

$$h_1 = 1/800, \tau_1 = 1/8000, \quad h_2 = h_1/2, \tau_2 = \tau_1/4.$$

Subscripts indicate matching parameters and solutions.

Let us briefly recall the dynamics of the development and destruction of oscillations from Section 4.3.3, which was studied in detail in [46]. After several periods of oscillations at some point in time (in the considered variant $\theta_{max}^{(1)} = 33.9$), the solution of the 'complete' problem leads to the formation of the first off-axis maximum of the electron density comparable in magnitude with the regular axial one. Then, nonlinear growth of this off-axis maximum occurs from period to period, leading to density inversion to infinity,

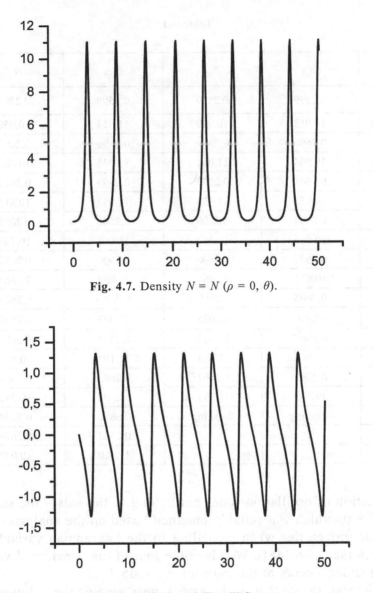

Fig. 4.7. Density $N = N\,(\rho = 0,\ \theta)$.

Fig. 4.8. Speed $W = W(\theta)$.

which corresponds to the formation of a jump (discontinuity) of the electric field. This effect is called «breaking», which essentially means a violation of the correctness of the hydrodynamic model of the process. The breaking time (i.e., the length of growth of the off-axis maximum) is determined by the parameters of the problem and can be from one to several hundred (and even thousands) periods. Fundamentally important is the fact that the spatial coordinate of the

Table 4.1

θ	$N_{f,1}$	$N_{f,2}$	$N_{a,2}$	$N_{a,*}$
00.0	0.2599	0.2599	0.2599	0.2599
03.0	10.913	10.990	11.014	10.958
05.9	0.2600	0.2599	0.2599	0.2607
08.9	10.934	11.012	11.035	10.885
11.8	0.2601	0.2599	0.2599	0.2615
14.8	10.952	11.032	11.053	10.816
17.7	0.2602	0.2600	0.2600	0.2623
20.7	10.966	11.049	11.068	10.745
23.6	0.2604	0.2600	0.2600	0.2632
26.6	10.975	11.064	11.082	10.665
29.5	0.2605	0.2600	0.2600	0.2641
32.5	10.980	11.075	11.093	10.568
35.4	0.2607	0.2601	0.2601	0.2651
38.4	10.980	11.083	11.101	10.446
41.3	0.2365	0.2602	0.2602	0.2661
44.3	39.893	11.088	11.107	10.292
45.7	Breaking	Breaking	0.6856	0.7598
47.2			0.2602	0.2674
50.2			11.110	10.097

destruction of oscillations does not belong to the axis – the straight line $\rho = 0$, which was reliably obtained based on the analysis of the particle trajectories when modelling in the Lagrangian variables in sec. 4.4 (see also [46]). We also note that in the considered variant the breaking occurs at the moment $\theta_{wb} = 45.7$.

It is easy to see that in the smoothness area of the solution (for $0 \le \theta \le 38.4$) the divergence of the values in the columns $Nf_{,1}$ and $N_{f,2}$ does not exceed 1%, however, in the vicinity of breaking oscillations (at $\theta \ge 41.3$), the differences are quite significant. This means that the grid parameters for solving the 'complete' problem were chosen quite satisfactory. Moreover, as shown in [45, 46], it is fundamentally impossible to reach the time θ_{wb} in calculations by the finite-difference method in Eulerian variables (this requires methods of the type of particles in Lagrangian variables). Therefore, we can assume that the column $N_{f,2}$ is a computationally correct solution to the 'complete' problem on the axis $\rho = 0$.

The content of the column $N_{a,2}$ corresponds to fig. 4.7 and demonstrates the values of the axial solution (electron density) at the times indicated in column θ and calculated for the same values of the parameters of the problem at $\tau = \tau_2$. A simple observation shows that the divergence at critical points (i.e., for extreme values) is no more than 0.22%, which indicates the correctness of the proposed approach to the calculation of axial solutions.

Let us pay attention to the main goal of the section – increasing the computational efficiency of algorithms aimed at obtaining a solution to the problem on the symmetry axis. In solving the 'complete' problem, calculations with parameters h_2 and τ_2 involve approximately $M = 4300$ points in space, and the leapfrog method is used to implement the scheme from [45]. This algorithm, of course, does not change the asymptotic behaviour of computational work, but it increases its constant K_p by about 3 times [83] due to additional calculations with variable matrix coefficients depending on the current solution. Therefore, with interest only in the axial behavior of the electron density, it is possible to achieve a reduction in the amount of calculations equal to $M \times K_p$, that is, more than 12,000 times.

Recall that such a reduction is achieved when the solutions practically coincide (the difference is of the order of 0.22%). If we allow a large error in relation to the solution of the 'complete' problem, the reduction will be even more significant. In the column $N_{a,*}$ table. 4.1 shows the results of calculations of the 'truncated' problem with a coarser time step ($\tau_* = 1/1000$). They correspond to approximately 10% (not distinguished visually in the graphical representation) errors: in this variant, the gain in the volume of calculations ($M \times K \times \tau_*/\tau_2$) is already more than 400 000 times.

Let us give another argument in favor of the reasonableness of computations of only axial solutions of problems: it has a more 'physical' than a 'mathematical' character. Solutions of 'complete' problems that simulate nonrelativistic cylindrical electron plasma oscillations always exist only on a limited time interval $\theta < \theta_{wb}$ (see [46, 131]). At the same time, reliable calculations of varying degrees of accuracy illustrate the regularity of axial solutions over longer time intervals. In particular, in the considered variant, the breaking occurs at the moment $\theta_{wb} = 45.7$, and the axial solutions are shown in Figs. 4.7 and 4.8 and in Table 4.1, including when $\theta > \theta_{wb}$. Thus, the conducted numerical simulation is a weighty additional argument in favor of the off-axis breaking of the considered oscillations of electrons in the plasma.

4.5.2. Forced relativistic oscillations

Considerably less work is devoted to the analysis of oscillations with regard for relativism than if this factor is neglected. Therefore, below we will be interested in numerical modelling of axial solutions of a somewhat more complex system of nonlinear equations, which are also generated not by the initial conditions, but by the right-hand side of a special form, forcing oscillations.

In this case, the formulation of the 'complete' problem is as follows. Consider the forced relativistic oscillations, which are described by the system of equations

$$\frac{\partial P}{\partial \theta} + E + \frac{\partial \gamma}{\partial \rho} = 0, \quad \gamma = \sqrt{1 + P^2 + |a|^2 / 2},$$

$$V = \frac{P}{\gamma}, \quad \frac{\partial E}{\partial \theta} - V\left[1 - \frac{1}{\rho}\frac{\partial}{\partial \rho}(\rho E)\right] = 0,$$

$$(4.57)$$

where the dimensionless required functions have the following meaning: $P(\rho, \theta)$ and $V(\rho, \theta)$ are the electron momentum and velocity, $E(\rho, \theta)$ is the electric field, $\gamma(\rho, \theta)$ is the relativistic (Lorentz) factor.

The physical interpretation of system (4.57) is as follows. A short superpowerful laser pulse, characterized by a given envelope $a(\rho, \theta)$, propagates along a cylindrical plasma bunch. The pulse speed is so high, and the length is so small that the reverse influence of the plasma on it can be neglected. The very effect of a laser pulse on a plasma is modeled by the dependence of the envelope on time and the transverse coordinate in the form

$$a(\rho, \theta) = a_* \exp\left\{-\frac{\rho^2}{\rho_*^2} - \frac{(\theta_{\min} - \theta)^2}{l_*^2}\right\}, \qquad (4.58)$$

where a_*, ρ_*, l_* are the given parameters. This formula means that at the initial time $\theta = 0$, the influence of the pulse is absent (for this, the $\theta_{\min}$ value must be taken sufficiently large, for example, for $l_* = 3.5$, $\theta_{\min}$ was assumed to be equal to 11 in the calculations). Further, as θ increases, the impact intensity first increases (up to $\theta = \theta_{\min}$

inclusive), then decreases at the same rate, and, starting from the moment $\theta \approx 2\,\theta_{min}$, its effect is almost absent. At the same time, at each moment of time, the intensity distribution over the transverse coordinate is Gaussian in character and rather rapidly decreases in space.

To uniquely determine the desired functions in the domain $\Omega = = \{(\rho,\theta):\ 0 \le \rho \le d,\ 0 \le \theta \le \theta_{max}$, it is necessary to add the boundary and initial conditions.

On the axis (at $\rho = 0$), due to the axial symmetry of the problem, we have

$$V(0,\theta)=P(0,\theta)=E(0,\theta)=0, \quad \gamma(0,\theta)=\sqrt{1+T^2(\theta)/2}, \qquad (4.59)$$

where the function $T(\theta)$ has the form

$$T(\theta)=a_* \exp\left\{-\frac{(\theta_{min}-\theta)^2}{l_*^2}\right\}.$$

In addition, the physical formulation assumes that the pulse cross section is substantially smaller than the transverse size of the region occupied by the plasma, that is, $\exp(-d^2/\rho_*^2)\ll 1$. Therefore, for a sufficiently large value of d, the equations

$$V(d,\theta)=P(d,\theta)=E(d,\theta)=0, \quad \gamma(d,\theta)=1. \qquad (4.60)$$

The initial data (we fix for convenience this moment as $\theta = 0$) correspond to the state of rest:

$$V(\rho,0)=P(\rho,0)=E(\rho,0)=0, \quad \gamma(\rho,0)=1. \qquad (4.61)$$

Recall that the expression (4.51) for the electron density

$$N(\rho,\theta)=1-\frac{1}{\rho}\frac{\partial}{\partial\rho}(\rho E).$$

Thus, taking into account relativistic effects, the problem of forced plasma oscillations is formulated as follows: find in the region Ω the functions V, E, P and γ, satisfying relations (4.57)–(4.61). As in the previous section, the peculiarity of the statement is preserved: of primary interest is the electron density function on the axis (at $\rho = 0$), but for this it is necessary to solve the 'complete' problem.

We describe the application of the finite difference method for solving the 'complete' problem from [100]. We introduce in the domain Ω a uniform grid with steps h and τ so that $\rho_m = mh$, $0 \le m \le M$, $Mh = d$ and $\theta_k = k\tau$, $0 \le k \le K$, $K\tau = \theta_{max}$. Using the notation $f_m^k = f(\rho_m, \theta_k)$ for grid functions, we write the discrete analog of equations (4.57):

$$\frac{P_m^{k+1} - P_m^k}{\tau} + E_m^k + \frac{\gamma_{m+1}^k - \gamma_{m-1}^k}{2h} = 0, \quad 1 \le m \le M-1,$$

$$\gamma_m^{k+1} = \sqrt{1 + \left(P_m^{k+1}\right)^2 + |a(\rho_m, \theta_{k+1})|^2 / 2},$$

$$V_m^{k+1} = \frac{P_m^{k+1}}{\gamma_m^{k+1}}, \quad 0 \le m \le M, \tag{4.62}$$

$$\frac{E_m^{k+1} - E_m^k}{\tau} - V_m^{k+1}\left[1 - \frac{1}{\rho_m} \frac{P_{m+1} E_{m+1}^{k+1} - P_{m-1} E_{m-1}^{k+1}}{2h}\right] = 0,$$

$$1 \le m \le M-1,$$

where $k \ge 0$. Difference initial and boundary conditions, as well as the expression for the envelope a, are defined in a natural way as the projections of relations (4.58)–(4.61) onto the grid.

The initial equations have non-linear terms, but the chosen scheme is non-iterative: the momentum P, the Lorentz factor γ and the velocity V are calculated using explicit formulas, and the electric field E requires the use of the tridiagonal matrix algorithm [22]. The properties of approximation and stability are somewhat unusual. In the nonlinear case for smooth solutions, the considered scheme has the order of approximation $O(\tau + h^2)$, and the principle of frozen coefficients [52] leads to the stability condition $\tau = O(h^2)$. For the linearized problem, the order of approximation in time rises to the second, and the scheme becomes unconditionally stable.

The grid analog of electron density N is calculated as follows:

$$N_m^k = \begin{cases} 1 - \dfrac{1}{\rho_m} \dfrac{\rho_{m+1}E_{m+1}^k - \rho_{m-1}E_{m-1}^k}{2h} & \text{at} \quad 1 \leqslant m \leqslant M-1, \\[3mm] 1 - 2\dfrac{E_1^k}{h} & \text{at} \quad m = 0, \\[3mm] 1 & \text{at} \quad m = M. \end{cases}$$

Consider the design variant according to the scheme (4.62) with the task parameters:

$$a_* = 5.0, \quad \rho_* = 5.0, \quad l_* = 3.5, \quad d = 22.5, \quad \theta_{\min} = 11.0.$$

In Table 4.2 at the time mements indicated in column θ, extreme values of the electron density N on the axis $\rho = 0$ are shown, that is, its maxima and minima. In this case, the values in the column $N_{f,1}$ were calculated with the grid parameters $h_1 = 2.5 \cdot 10^{-3}$, $\tau_1 = 0.3125 \cdot 10^{-4}$, and the values of $N_{f,2}$ – with $h_2 = h_1/2$, $\tau_2 = \tau_1/4$. In this case, the grid parameters were chosen so small that the breaking of grid solutions was as synchronous as possible.

It is easy to see that here the differences in the extreme values of $N_{f,1}$ and $N_{f,2}$ do not exceed 0.03%, which indicates a high reliability of the calculations, and not only in the smoothness of the solution. For the taken physical parameters of the problem, the moment of breaking (density becomes infinite) is defined as $\theta_{wb} \approx 37.5$, therefore, starting from this time, the solution to the 'complete' problem (4.57)–(4.61) is generally not defined.

Consider the formulation of a 'truncated' problem. Taking into account the form of the boundary conditions (4.59), we will consider in the vicinity of the axis linear solutions in space P, V, E of equations (4.57), i.e.

$$V(\rho,\theta) = W(\theta)\rho, \quad E(\rho,\theta) = D(\theta)\rho, \quad P(\rho,\theta) = Q(\theta)\rho, \qquad (4.63)$$

and the influence of the relativistic factor we consider as follows. First consider the full expression

$$\frac{\partial \gamma}{\partial \rho} = \frac{1}{2} \frac{1}{\sqrt{1+P^2+|a|^2/2}} \left(2P\frac{\partial P}{\partial \rho} + \frac{1}{2}\frac{\partial |a|^2}{\partial \rho} \right), \qquad (4.64)$$

Table 4.2

θ	$N_{f,1}$	$N_{f,2}$	$N_{a,2}$	$N_{a,*}$
14.4	0.3497	0.3497	0.3497	0.3497
17.7	5.2155	5.2158	5.2158	5.1790
20.8	0.3425	0.3425	0.3425	0.3448
23.8	5.2453	5.2460	5.2460	5.1262
26.8	0.3427	0.3427	0.3427	0.3473
29.9	5.2406	5.2416	5.2417	5.0486
32.9	0.3425	0.3425	0.3425	0.3493
35.9	5.2266	5.2282	5.2283	4.9443
39.0	Breaking	Breaking	0.3426	0.3514
42.0			5.2469	4.8951
45.0			0.3427	0.3542
48.1			5.2376	4.8368

in which we calculate

$$\frac{\partial |a|^2}{\partial \rho} = -\frac{4\rho}{\rho_*^2}\exp\left\{-\frac{2\rho^2}{\rho_*^2}\right\}T^2(\theta).$$

Now, analyzing in (4.64) the linear dependence for small ρ, taking into account (4.63), we obtain

$$\frac{\partial \gamma}{\partial \rho} = \frac{1}{\sqrt{1+T^2(\theta)/2}}\left(Q^2 - \frac{T^2(\theta)}{\rho_*^2}\right)\rho + o(\rho).$$

In addition, from (4.57) we have

$$V\gamma = P, \quad \gamma = \sqrt{1+P^2+|a|^2/2},$$

what under the assumption (4.63) gives

$$\Gamma(\theta) \equiv \gamma(\rho=0,\theta) = \sqrt{1+T^2(\theta)/2}, \quad W(\theta) = Q(\theta)/\Gamma(\theta).$$

As a result, for axial solutions we have a system of differential and algebraic equations

$$Q' + D + \frac{Q^2}{\Gamma(\theta)} = \frac{T^2(\theta)}{\Gamma(\theta)\rho_*^2}, \quad W = \frac{Q}{\Gamma(\theta)}, \quad D' - W(1-2D) = 0 \qquad (4.65)$$

with known (specified) functions $\Gamma(\theta)$, $T(\theta)$, supplemented by the initial conditions of rest at $\theta = 0$:

$$W(0) - Q(0) - D(0) - 0 \qquad (4.66)$$

Recall that, similar to the previous section (where nonrelativistic oscillations are considered) for the electron density function there is an expression on the axis

$$N(\rho = 0, \theta) = 1 - 2D(\theta).$$

Thus, the problem of determining axial solutions in this case reduces to the integration of equations (4.65) on the time interval $0 \ \square \ \theta \ \square \ \theta_{max}$, starting from the initial conditions (4.66).

We describe the numerical scheme and the results of calculations for the 'truncated' problem. We introduce a uniform grid $\theta_k = k\tau$, $0 \le k \le K$, $K\tau = \theta_{max}$ and write the discrete analog of equations (4.65). Using the notation $f_k = f(\theta_k)$ for grid functions, we will have for $k \ge 0$ the scheme

$$\frac{Q_{k+1} - Q_k}{\tau} + D_k + \frac{Q_k^2}{\Gamma(\theta_k)} = \frac{T^2(\theta_k)}{\Gamma(\theta_k)\rho_*^2},$$

$$W_{k+1} = \frac{Q_{k+1}}{\Gamma(\theta_{k+1})}, \quad \frac{D_{k+1} - D_k}{\tau} - W_{k+1}(1 - 2D_{k+1}) = 0. \qquad (4.67)$$

Difference initial conditions are determined in a natural way:

$$W_0 = Q_0 = D_0 = 0.$$

It is easy to note that for $\theta \ge 2\,\theta_{min}$ (when $\Gamma(\theta) \approx 1$, $T(\theta) \approx 0$), relations (4.67) go into relations (4.56), since $Q(\theta) \approx W(\theta)$. Therefore, when forced oscillations are transformed into free ones, the properties of the circuits (4.67) and (4.56) completely coincide.

For comparison with previous calculations, we present the results of calculations according to the scheme (4.67) with the same parameters of the problem. Consider in table. 4.2 column

$N_{a,\,2}$, containing the values of electron density, calculated with the grid parameter τ_2. It is easy to see that the extreme values on the axis obtained from the 'complete' and 'truncated' tasks with the same physical and grid parameters almost coincide (there are differences in the last digits for some maxima).

Taking into account the dimensions of the above tasks and additional computational work related to the tridiagonal matrix algorithm in the case of variable coefficients for the scheme (4.62), in this case we have a reduction in the volume of calculations ($M \times K_p$) by more than 50 000 times with an error not exceeding $2 \cdot 10^{-3}\%$. If an error of the order of 10% is allowed (not distinguishable in a graphical representation), then (see the column $N_{a,*}$ in Table 4.2), it is enough to take $\tau_* = 1/200$ instead of τ_2 to calculate the density values only on the axis reduce the volume of computational work ($M \times K_p \times \tau_* / \tau_2$) by more than $8 \cdot 10^6$ times compared to solving the complete problem, from which the column $N_{f,2}$ is obtained.

We note that the columns in which the calculations of the 'truncated' problem are given, with the same parameters, continue further in time. This means that problem (4.65), (4.66) has a regular solution even for $\theta > \theta_{wb}$. In other words, the above calculations indicate an off-axis breaking of forced relativistic oscillations. Additional illustrations to this fact are in Figs. 4.9 and 4.10. In Fig. Fig. 4.9 is a graph of the electron density obtained by solving the 'truncated' problem at times of $0 \le \theta \le 50$, which significantly exceed the breaking time. In fig. 4.10 shows the time dependence of $\max\limits_{0 \le \rho \le d} N(\rho,\,\theta)$, obtained by calculating the 'complete' problem. The coincidence of the first three global maxima with local (axial) is well noticeable. Next, the first in time ($\theta_{max}^{(1)} \approx 32$) off-axis maximum is formed with a value approximately twice the axial regular maxima. The growth of precisely this maximum leads to the breaking of oscillations at the moment θ_{wb}.

Let us summarize the results of this section. For cylindrical oscillations, 'truncated' problems are obtained that describe the behaviour of the solution on the axis of symmetry. These dedicated solutions are called, for convenience, axial. For 'truncated' problems, numerical algorithms are considered and versatile computational experiments are conducted. They, in particular, imply the correctness (validity) of the derived one-dimensional systems of equations. We note that in the case of free nonrelativistic and forced relativistic oscillations, the 'truncated' equations are exact. In addition, theoretical estimates are given below (in practice, they

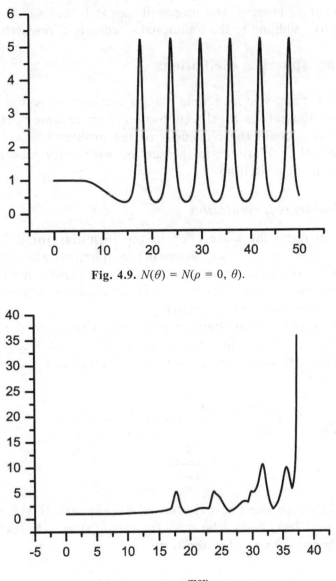

Fig. 4.9. $N(\theta) = N(\rho = 0, \theta)$.

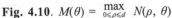

Fig. 4.10. $M(\theta) = \max\limits_{0 \leqslant \rho \leqslant d} N(\rho, \theta)$

are significantly larger) reducing the volume of computational work when calculating only axial solutions as compared to determining the solution to the 'complete' problem. And, finally, a side effect of the simulation is confirmation of the fact of off-axis breaking of oscillations excited by a sharply focused laser pulse in a plasma.

Of course, it should always be borne in mind that modelling only axial solutions may not be sufficient for a detailed study of a phenomenon or process. The proposed approach seems to be a useful and effective addition to the traditional – 'complete' research scheme.

4.6. About spherical oscillations

Spherical symmetry in modelling plasma oscillations, both theoretical and experimental, is rarely encountered in research. However, taking into account the closeness of the problems of cylindrical and spherical oscillations, in the section we briefly present useful information for modelling.

4.6.1. Problems formulation

Bearing in mind the difference in the formulas for differential operators (curl, divergence, gradient) in spherical and cylindrical coordinates, from the basic equations of the plasma model under consideration (1.5)–(1.12) it is easy to obtain a system whose solutions have spherical symmetry.

Omitting the intermediate calculations, similar to the ones in section 4.1, we write the equations describing the free spherical one-dimensional relativistic oscillatory motion of electrons in a cold ideal plasma:

$$\frac{\partial P}{\partial \theta} + E + V\frac{\partial P}{\partial \rho} = 0, \quad \frac{\partial E}{\partial \theta} - V + \frac{V}{\rho^2}\frac{\partial}{\partial \rho}\left(\rho^2 E\right) = 0,$$

$$V = \frac{P}{\sqrt{1 + P^2}}.$$

(4.68)

Let us introduce the abbreviation S1RE (Spherical 1-dimension Relativistic Electron oscillations) for this system. The expression for the electron density has the form

$$N = 1 - \frac{1}{\rho^2}\frac{\partial}{\partial \rho}\left(\rho^2\right).$$

(4.69)

Taking into account the specificity of the spherical coordinate system ($\rho \geq 0$), we will study the solutions of the S1RE system in the $\{(\rho, \theta)$ semi-infinite strip: $0 < \rho < d, \theta >> 0\}$. The introduction of the parameter d, limiting the spatial domain, is based on the results of previous chapters. Equation of equations (4.68) with initial conditions localized in space

$$P(\rho,0) = P_0(\rho), \quad E(\rho,0) = E_0(\rho), \quad \rho \in [0,d], \tag{4.70}$$

and the corresponding boundary conditions

$$P(0,\theta) = P(d,\theta) = E(0,\theta) = E(d,\theta) = 0 \quad \forall \theta \geqslant 0. \tag{4.71}$$

From where, in particular, using the algebraic relation connecting the velocity V and the momentum P of electrons, the equalities follow

$$V(0,\theta) = V(d,\theta) = 0 \quad \forall \theta \geqslant 0,$$

not allowing the transfer of perturbations of the unknown functions across the boundaries of the segment $[0, d]$.

As before, we will keep in mind the following functions as initial conditions (4.70):

$$P_0(\rho) = 0, \quad E_0(\rho) = \begin{cases} \left(\dfrac{a_*}{\rho_*}\right)^2 \rho \exp^2\left\{-\dfrac{\rho^2}{\rho_*^2}\right\}, & 0 \leqslant \rho < d, \\ 0, & \rho = d. \end{cases} \tag{4.72}$$

Such a perturbation of the electric field, initiating oscillations, is taken by analogy with the formulation of the problem for cylindrical oscillations. By virtue of the exponential dumping of the function $E_0(\rho)$, in order to ensure with sufficient accuracy the boundary condition (4.71), it suffices to set $d = 4.5\rho_*$.

To compare the presence/absence of effects of relativism, a simpler system of equations is required. We assume that the electron velocity is essentially nonrelativistic, i.e.

$$P \approx V, \quad \frac{\partial P}{\partial \rho} \approx \frac{\partial V}{\partial \rho}, \quad \frac{\partial P}{\partial \theta} \approx \frac{\partial V}{\partial \theta}.$$

In this case, we arrive at the equations describing the free cylindrical one-dimensional non-relativistic oscillatory movements of electrons in a cold ideal plasma:

$$\frac{\partial V}{\partial \theta} + E + V\frac{\partial V}{\partial \rho} = 0, \quad \frac{\partial E}{\partial \theta} - V + \frac{V}{\rho^2}\frac{\partial}{\partial \rho}\left(\rho^2 E\right) = 0. \tag{4.73}$$

Let's designate it as S1NE (Spherical 1-dimension Nonrelativistic Electron oscillations).

Similar to the relativistic case, we assume that it is required in the semi-infinite strip $\{(\rho, \theta): 0 < \rho < d, \theta > 0\}$ to find the solution of equations (4.73) that satisfies the local initial

$$V(\rho,0)=V_0(\rho), \quad E(\rho,0)=E_0(\rho), \quad \rho \in [0,d], \tag{4.74}$$

and marginal conditions

$$V(0,\theta)=V(d,\theta)=E(0,\theta)=E(d,\theta)=0 \quad \forall \theta \geqslant 0. \tag{4.75}$$

In this case, we will keep in mind that $V_0(\rho) = 0$, and the function $E_0(\rho)$ is the same in (4.70) and (4.74).

The quasilinear system of equations S1NE (4.73) is important for the construction of numerical algorithms, therefore, in addition to writing in the Eulerian variables, its form in the Lagrangian variables will be useful:

$$\frac{dV(\rho^L,\theta)}{d\theta}=-E(\rho^L,\theta),$$

$$\frac{dE(\rho^L,\theta)}{d\theta}+2\frac{E(\rho^L,\theta)V(\rho^L,\theta)}{\rho}=V(\rho^L,\theta), \tag{4.76}$$

where $d/d\theta = \partial/\partial\theta + V\partial/\partial\rho$ is the total time derivative.

Recall that the function $R(\rho^L, \theta)$, which determines the displacement of a particle with the Lagrangian coordinate ρ^L, so that

$$\rho(\rho^L,\theta)=\rho^L+R(\rho^L,\theta), \tag{4.77}$$

satisfies the equation

$$\frac{dR(\rho^L,\theta)}{d\theta}=V(\rho^L,\theta). \tag{4.78}$$

Expressing the velocity V through the offset R in accordance with formula (4.78), we write the second equation (4.76) in the form

$$(\rho^L + R)^2 \frac{dE}{d\theta} + 2(\rho^L + R)E\frac{dR}{d\theta} = (\rho^L + R)^2 \frac{dR}{d\theta}. \qquad (4.79)$$

Equation (4.79) has the first integral

$$(\rho^L + R)^2 E = \frac{1}{3}(\rho^L + R)^3 + C,$$

where the constant C is determined from the condition that the electric field is zero in the absence of particle displacement. Then from this relation we find the expression for the electric field

$$E(\rho^L, \theta) = \frac{1}{3}\frac{(\rho^L + R(\rho^L, \theta))^3 - (\rho^L)^3}{(\rho^L + R(\rho^L, \theta))^2}, \qquad (4.80)$$

and the basic system of equations (4.73) in Lagrangian variables takes the form

$$\frac{dV}{d\theta} = -E, \quad \frac{dR}{d\theta} = V. \qquad (4.81)$$

We note that relation (4.77) is very useful for determining the Lagrangian coordinate of the particle ρ^L and the initial condition $R(\rho^L, 0)$ from the given distribution $E_0(\rho)$. The algorithm is as follows: for some ρ from the equation (see the ratio (4.80) with $\theta = 0$)

$$\frac{1}{3}\frac{\rho^3 - (\rho^L)^3}{\rho^2} = E_0(\rho)$$

by the explicit formula is determined by the value

$$\rho^L = \left(\rho^3 - 3\rho^2 E_0(\rho)\right)^{1/3}, \qquad (4.82)$$

and then from (4.77) is the desired initial displacement at the point ρ^L:

$$R(\rho^L, 0) = \rho - \rho^L. \qquad (4.83)$$

To summarize the above, the trajectories of all particles, each of which is identified by the Lagrangian coordinate ρ^L, can be determined by independent integration of the system of ordinary differential equations (4.80), (4.81). For this, two initial conditions are required: $R(\rho^L, 0)$ and $V(\rho^L, 0)$. From (4.72), (4.74) we have $V(\rho^L, 0) = 0$. To define $R(\rho^L, 0)$, first set the position of the particle ρ at the initial time, then the Lagrangian coordinate is determined by the formula (4.82), and the initial offset is calculated according to the formula (4.83). Knowledge of the Lagrangian coordinate ρ^L and the function of displacement $R(\rho^L, \theta)$ uniquely characterizes the trajectory of a particle by the formula (4.77).

At the end of the section, it should be noted that for the relativistic system of equations S1RE (4.68), its analogue in Lagrangian variables has the form

$$\frac{dP(\rho^L,\theta)}{d\theta} = -E(\rho^L,\theta),$$

$$\frac{dR(\rho^L,\theta)}{d\theta} = \frac{P(\rho^L,\theta)}{\sqrt{1+P^2(\rho^L,\theta)}} \equiv V(\rho^L,\theta), \tag{4.84}$$

the expression for the electric field (4.80) and, therefore, the method of specifying the initial conditions ρ^L, $R(\rho^L, 0)$ and $P(\rho^L, 0) = V(\rho^L, 0) = = 0$ remain unchanged.

4.6.2. Axial solution

The axial solution of the equations S1NE (4.73) is understood as a real solution of the form

$$V(\rho,\theta) = W(\theta)\rho, \quad E(\rho,\theta) = D(\theta)\rho.$$

It is easy to make sure that the time-dependent factors in this case satisfy the system of ordinary differential equations.

$$W' + D + W^2 = 0, \quad D' - W + 3WD = 0. \tag{4.85}$$

We supplement the equations obtained with arbitrary real initial conditions

$$W(0) = \beta, \quad D(0) = \alpha \tag{4.86}$$

and proceed to clarify the condition for the existence and uniqueness of the solution of the Cauchy problem (4.85), (4.86).

Note that the reduced Cauchy problem is not trivial, since it admits both regular periodic solutions (for example, for small α and β) and solutions that have singularities on a finite time interval (the so-called blow-up solutions). Therefore, it seems useful to have a slightly different look at the considered problem. Takes place

Lemma 4.6.1. *The Cauchy problem* (4.85), (4.86) *is equivalent to the following differential algebraic problem:*

$$W' + (\beta^2 - 2\alpha + 1)x^2 + (3\alpha - 1)x^3 = 0, \tag{4.87}$$

$$3W^2 + 1 - 3(\beta^2 - 2\alpha + 1)x^2 - 2(3\alpha - 1)x^3 = 0, \tag{4.88}$$

$$W(0) = \beta, \quad x(0) = 1. \tag{4.89}$$

Proof. After excluding the function D from system (4.85), we obtain the Cauchy problem for a second-order equation:

$$W'' + 5W'W + W + 3W^3 = 0, \quad W(0) = \beta,$$
$$W'(0) = -(\alpha + \beta^2). \tag{4.90}$$

Reduce the order of the equation by replacing $p(W) = W'_\theta$:

$$p'_W p + 5pW + W + 3W^3 = 0. \tag{4.91}$$

Hereinafter, the subscript of the derivative clearly indicates the independent variable by which differentiation is carried out. Note that equation (4.91) corresponds (see problem (4.90)) to the initial condition

$$p(\beta) = -(\alpha + \beta^2). \tag{4.92}$$

The transformation of the dependent variable $p(W) = u^{-1}(W) \neq 0$ leads to the equation

$$u'_W - 5u^2 W - u^3(W + 3W^3) = 0,$$

in which it is convenient to make the substitution $u(W) = \eta(\xi)$, where $\xi = \frac{5}{2}W^2 + C_\xi$. As a result, we will have

$$\eta'_\xi = g(\xi)\eta^3 + \eta^2, \quad g(\xi) = \frac{6}{25}\xi + \frac{6}{25}\left(\frac{5}{6} - C_\xi\right).$$

To obtain an analytical solution of this equation, we introduce the parameterization of the independent variable $\xi = \xi(t)$ so that

$$\xi'_t = -\frac{1}{t\eta(\xi)}, \quad t \neq 0,$$

and as its corollary we come to the equation

$$t^2 \xi''_t + \frac{6}{25}\xi + \frac{6}{25}\left(\frac{5}{6} - C_\xi\right) = 0.$$

Its general solution is

$$\xi(t) = C_1 t^{3/5} + C_2 t^{2/5} + C_\xi - \frac{5}{6},$$

where does the formula come from

$$\eta(\xi) = -\left(\frac{3}{5}C_1 t^{3/5} + \frac{2}{5}C_2 t^{2/5}\right)^{-1}.$$

Returning to the original variables gives

$$p(W) = -\left(\frac{3}{5}C_1 t^{3/5} + \frac{2}{5}C_2 t^{2/5}\right), \quad \frac{5}{2}W^2 + \frac{5}{6} = C_1 t^{3/5} + C_2 t^{2/5}.$$

The transformations leading to these relations are completely analogous to those described in detail in the proof of Lemma 2.2.1.

The definition of the constants C_1, C_2 from condition (4.92), that is, from the agreement of the values of the parameters $\theta = 0$ and t

= 1, and the formal replacement of $t^{1/5} = x$ lead to the differential-algebraic problem (4.87)–(4.89). The lemma is proved.

It should be clarified that in this case the equivalence of the considered statements is understood as follows: the function $W(\theta)$ in both formulations is the same, and the remaining functions $D(\theta)$ and $x(\theta)$ are determined in each formulation by $W(\theta)$ uniquely.

For the case of spherical oscillations, an analogue of Theorem 4.2.1 has not yet been established. In particular, this is due to the rather complex structure of the manifold (4.88), which is an elliptic curve [72]. However, the following hypothesis seems plausible.

Hypothesis. *A necessary and sufficient condition for the existence and uniqueness of a smooth periodic solution of the Cauchy problem* (4.85), (4.86) *is the fulfillment of the inequality*

$$\alpha < 1/3.$$

The necessity of this condition can be established by reasoning as in similar situations in Theorems 2.2.1 and 4.2.1.

4.6.3. Perturbation method

If we assume that the oscillation amplitude is sufficiently small, that is, in the initial condition (4.72) for the electric field $a_* \ \Box\ \rho_*$, then the equations S1NE (4.73) become weakly nonlinear and their approximate solutions can be constructed by analogy with cylindrical oscillations, using perturbation theory methodology [25, 154].

Eliminating the electric field E from the system (4.73), we obtain the following equation for the velocity V_p:

$$\left(\frac{\partial^2}{\partial\theta^2}+1\right)V_p + \frac{\partial}{\partial\theta}\left(V_p\frac{\partial V_p}{\partial\rho}\right) + V_p\frac{1}{\rho^2}\frac{\partial}{\partial\rho}\left(\rho^2\frac{\partial V_p}{\partial\theta}\right) +$$
$$+V_p\frac{1}{\rho^2}\frac{\partial}{\partial\rho}\left(\rho^2 V_p\frac{\partial V_p}{\partial\rho}\right) = 0. \tag{4.93}$$

Here, the subscript p indicates approximations sought in accordance with perturbation theory.

Equation (4.93) is solved taking into account the smallness of nonlinear terms. Substituting the velocity of electrons in the form of expansion in powers of nonlinearity

$$V_p = V_1 + V_2 + V_3 + \dots,$$

we find that the first approximation satisfying the initial conditions

$$V_1\big|_{\theta=0} = 0, \quad \frac{\partial V_1}{\partial \theta}\bigg|_{\theta=0} = A(\rho),$$

has the appearance

$$V_1 = A\sin\theta, \tag{4.94}$$

where $A = A(\rho)$ is the amplitude of the electron velocity oscillations depending on the radius.

For the second and third terms of the asymptotic expansion of the solution (4.93), that is, V_2 and V_3, equations are successively derived in the usual way, which can be omitted here without losing meaning.

Further, to eliminate the resonant terms in the equation for the third term V_3, which lead to solutions increasing in time, the frequency change in the formula (4.94) should be taken into account:

$$V_1 = A\sin\omega\theta, \quad \omega = 1 + \omega_2, \tag{4.95}$$

moreover, ω_2 is a quadratic in amplitude correction to the fundamental frequency. As a result, we get

$$\omega_2 = \frac{7A^2(\rho)}{48\rho^2}. \tag{4.96}$$

This expression is not new, as it was already encountered in [131]. In this case, of greater interest is the formula for the electron density, similar to the cylindrical case from § 4.2.4, which follows from expression (4.95) and approximation $V_p \approx V_1$. Such a representation means that from the asymptotic representation of the velocity one should keep only the first term, although the nonlinear frequency shift (4.37) follows from the regular behavior of the three terms in the expansion in a small parameter of the oscillation amplitude.

For the S1RE relativistic equations, a natural way (see [46]) is followed by a refinement for the quadratic frequency shift:

$$\omega_2 = \frac{7A^2(\rho)}{48\rho^2} - \frac{3A^2(\rho)}{16}.$$

4.6.4. For numerical modelling

The construction of numerical algorithms for modelling spherical oscillations has no fundamental differences compared with the cylindrical case. Therefore, we note below only the formulas of practical interest.

Finite difference method. In the system of equations S1NE (1.73), in essence, the interaction of two physical processes is presented: nonlinear oscillations at a fixed point in space and their space-time transfer.

We present equations (4.73) to a convenient form. To this end, we single out the explicit term in the second equation, which is responsible for the shift of the oscillation frequency. As a result, we get

$$\frac{\partial V}{\partial \theta} + E + V\frac{\partial V}{\partial \rho} = 0, \quad \frac{\partial E}{\partial \theta} - V + 2\frac{VE}{\rho} + V\frac{\partial E}{\partial \rho} = 0.$$

We now relate to the description of the process of nonlinear oscillations of the equation

$$\frac{\partial \tilde{V}}{\partial \theta} + \tilde{E} = 0, \quad \frac{\partial \tilde{E}}{\partial \theta} - \tilde{V} + 2\frac{\tilde{V}\tilde{E}}{\rho} = 0, \tag{4.97}$$

and to their transfer in space and time –

$$\frac{\partial \bar{V}}{\partial \theta} + \bar{V}\frac{\partial \bar{V}}{\partial \rho} = 0, \quad \frac{\partial \bar{E}}{\partial \theta} + \bar{V}\frac{\partial \bar{E}}{\partial \rho} = 0. \tag{4.98}$$

As a basis for the time sampling of both systems, we use the usual leapfrog scheme. Let τ be a time step, then we will assign the values E, $\tilde{E}$, $\bar{E}$, N to 'integer' time moments $\theta_j = j\tau$ ($j \geq 0$ – integer) and $\theta_{j\pm1-2}$ – V, $\tilde{V}$, $\bar{V}$ to 'half integer'. The choice of the corresponding time moment for the function value will be denoted by superscript. For discretization in space, we will use a grid with a constant step h so that $\rho_m = mh$, $0 \leq m \leq M$, $Mh = d$.

We write the difference equations approximating the system (4.97):

$$\frac{\tilde{V}_m^{j+1/2} - \tilde{V}_m^{j-1/2}}{\tau} + \tilde{E}_m^j = 0,$$

$$\frac{\tilde{E}_m^{j+1} - \tilde{E}_m^j}{\tau} - \tilde{V}_m^{j+1/2} + \frac{\tilde{V}_m^{j+1/2}}{\rho_m}\left[\tilde{E}_m^{j+1} + \tilde{E}_m^j\right] = 0, \qquad (4.99)$$

$$\tilde{V}_m^{j-1/2} = V_m^{j-1/2}, \quad \tilde{E}_m^j = E_m^j, \quad 1 \leqslant m \leqslant M-1.$$

The difference from the difference approximation of the cylindrical equations (4.43) is only in the second equation. Difference approximations for the transport equations (4.98) do not differ from the approximations (4.44), therefore they are not given here.

Particle method. Here, the differences from the modelling of cylindrical vibrations are even smaller. To determine the trajectories of particles in Lagrangian variables, the former equations are used

$$\frac{dR}{d\theta} = V, \quad \frac{dV}{d\theta} = -E. \qquad (4.100)$$

However, for their closure in the spherical case, a slightly different formula is used, which explicitly describes the connection between the electric field and the displacement:

$$E = \frac{1}{3}\frac{\left(\rho^L + R\right)^3 - \left(\rho^L\right)^3}{\left(\rho^L + R\right)^2}. \qquad (4.101)$$

As a consequence, attention should be paid to the determination of the initial positions of the particles, which create a given electric field.

Suppose that in the initial moment of time $\theta = 0$, the particle with the number k is characterized by the initial position along the radius $\rho_0(k)$ and the initial displacement $R(k, 0)$, where $1 \leq k \leq M$, M is the total number of particles. The initial positions of all particles form the electric field, which has the form (4.72). On the other hand, the deviation of particles at the initial moment of time creates at the point with the coordinate $\rho_k = \rho_0(k) + R(k, 0)$ an electric field in accordance with formula (4.101). Comparing expressions (4.72) and (4.101), we can determine the desired values of $\rho_0(k)$ and $R(k, 0)$. To do this, we define the initial spatial grid $\rho_k = kh$, h is the discretization parameter with respect to the radial variable, which characterizes the proximity of neighboring particles. In grid nodes, using formula (4.72), we calculate the values of the electric field E $(\rho_k, 0)$. This electric field is formed by the displacement of particles, that is, on the basis of (4.101) we have the equations for determining the initial positions $\rho_0(k)$:

$$E(\rho_k, 0) = \frac{1}{3} \frac{\rho_k^3 - \rho_0^3(k)}{\rho_k^2},$$

and then, recalling that $\rho_k = \rho_0(k) + R(k, 0)$, from the already found initial positions of the particles, we find their initial deviations $R(k, 0)$ from these positions. Thus, to calculate the trajectory of each particle, initial data arc obtaincd, to which wc must add the condition of particle immobility at the initial time instant from (4.72), that is, $V(k, 0) = 0$.

Numerical experiments carried out both by the finite difference method and the particle method showed the presence of the effect of breaking plasma oscillations at a certain distance from the point $\rho = 0$. A detailed study of spherical oscillations, for example, for the presence of any new effects, has not yet been carried out.

4.7. Bibliography and comments

The results of this chapter are important as a kind of «bridge» to the chapters, in which the modelling of both two-dimensional oscillations close to axisymmetric waves and wake waves excited by a short laser pulse is presented. The fact is that cylindrical electron oscillations, initiated even by a weak laser pulse, still end up with a breaking effect. Therefore, in the methodical outline, the study of cylindrical oscillations is very useful to conduct before modelling axially symmetric wake waves in a plasma.

Although in historical terms, it was the opposite. A small research team led by an outstanding physicist L.M. Approximately from the end of the 80s of the last century, Gorbunov began to study the wake waves in a plasma excited by a focused laser pulse. This was connected with the most 'modern' scientific theme in plasma physics of that time — the acceleration of electrons using a wake wave [44, 166]. For some time now, the point of view dominated in this scientific field that the wake wave breaks on the axis of a laser pulse due to its peculiarity in the 'dovetail' type solution [119]. Our calculations showed that the breaking is fundamentally off-axis in nature, and, first of all, it concerns the keel of the watering waves, initiated by not very powerful pulses [109, 125, 126, 136]. Here, the breaking effect is usually preceded by the formation of the off-axis maximum of the electron density function. Further, a strongly nonlinear growth of this maximum leads to a singularity of the electron density, while on the axis of symmetry the regular density maxima are almost equal to.

Therefore, in order to accurately substantiate the effect of off-axis breaking, a number of studies were carried out, primarily related to the breaking of axially symmetric electron oscillations [45, 46, 123], and then a close relationship was observed between the breaking of these oscillations and wake plasma waves [93]. It should be noted that the work on the study of axial solutions [79, 100] are directly related to this rationale. At the present time, the possibility of off-axis breaking of axially symmetric electronic oscillations and wake waves excited by a laser pulse is not questioned. It was established that the initial point of view, called «Transverse breaking», was based primarily on numerical experiments that were not very carefully carried out due to the limited computational resources used in those ancient times.

In the given chapter, the nonrelativistic equations C1NE were studied in detail, since it was their analysis that made it possible to arrive at the desired, fundamentally important conclusions without unnecessarily complicating the model. On the other hand, studies of the relativistic C1RE equations were carried out in Refs [46, 123]. They showed that the shift of the oscillation frequency depends on both geometrical and relativistic factors:

$$\omega_2 = \frac{A^2(\rho)}{12\rho^2} - \frac{3A^2(\rho)}{16},$$

however, this does not significantly affect the nature of breaking. The numerical algorithms given in the chapter, which are constructed both by the finite difference method and the particle method, are easily transferred to the relativistic model. Moreover, the conducted numerical experiments [46] convincingly testified to the weak influence of relativism for small ρ_*. It should also be noted that the axial solutions of nonrelativistic and relativistic equations coincide. All of the above means that it is impractical in this case to clutter the presentation with meaningless formulas and calculations related to relativism.

It should be clarified that for the numerical analysis of the effect of the breaking of cylindrical oscillations only the algorithm based on Lagrangian variables is sufficient. Moreover, the calculated particle trajectories actually generate at each instant a non-uniform Eulerian grid, most suitable for describing (plotting) the electron density function. Therefore, a scheme based on the Eulerian variables was used primarily to control Lagrangian calculations. However, there are a number of more complex problems associated with both flat and cylindrical (spherical) electron oscillations, for example, taking into account the effects of ionization and recombination, viscosity, resistance and dissipativity of the plasma (see [141, 143, 161, 169]). The list should include tasks that take into account the movement of ions. For such statements, the construction of algorithms based on the Lagrangian variables becomes difficult, while the Eulerian methods allow a natural simple generalization to these listed and close to them statements.

The main results of the chapter were obtained in [45, 46, 79, 100, 123].

Spherical plasma oscillations devoted extremely few publications. A nonrelativistic formulation was considered in [131], in which a quadratic frequency shift was first established depending on the amplitude

$$\omega_2 = \frac{7A^2(\rho)}{48\rho^2}.$$

Relativistic effects were discussed in [117], although the article was not without curiosity, namely: for spherically symmetric oscillations, the authors 'found' that the frequency shift there is determined by the formula for the frequency shift with cylindrically symmetric oscillations in the plasma.

5

Influence of ion dynamics on plane one-dimensional oscillations

In the chapter, the influence of ion dynamics on the breaking of plane relativistic one-dimensional electron oscillations is studied using numerical simulation methods. When using the finite difference method, a computational algorithm based on the Eulerian variables is constructed. A preliminary analytical study of the problem has been carried out, and a new type of breaking of long-lived oscillations, different from those considered in previous chapters, has been presented.

5.1. Formulation of the problem

A preliminary study of plasma phenomena, as a rule, begins with a mathematical model in which ions are considered immobile. This approach is based on the fact that the masses of an electron and a proton differ by more than three orders of magnitude (the mass ratio is approximately equal to 1836), which makes the dynamics of ions hardly noticeable, especially against the background of high-speed movements of the electrons. However, when the processes under consideration continue for a rather long time (for example, multi-period plasma oscillations), the influence of even small quantities can lead to a qualitative change in the objects of observation.

In order to analyze the influence of ion dynamics on one-dimensional plane plasma oscillations, the main system (1.5)–(1.12) can be significantly simplified. Let us assume that

- the solution is determined only by the x-components of the vector functions $\mathbf{p}_e$, $\mathbf{v}_e$, $\mathbf{v}_i$ and $\mathbf{E}$, we denote them by p_e, v_e, v_i and E_x, respectively;
- there is no dependence in the above functions on the variables y and z, that is, $\partial/\partial y = \partial/\partial z = 0$.

Then from the system (1.5)–(1.12) we have

$$\frac{\partial n_e}{\partial t} + \frac{\partial}{\partial x}(n_e v_e) = 0, \quad \frac{\partial p_e}{\partial t} = eE_x - m_e c^2 \frac{\partial \gamma}{\partial x}, \quad \gamma = \sqrt{1 + \frac{p_e^2}{m_e^2 c^2}},$$

$$v_e = \frac{p_e}{m_e \gamma}, \quad \frac{\partial n_i}{\partial t} + \frac{\partial}{\partial x}(n_i v_i) = 0, \quad \frac{\partial v_i}{\partial t} + v_i \frac{\partial v_i}{\partial x} = \frac{e_i}{m_i} E_x, \tag{5.1}$$

$$\frac{\partial E_x}{\partial t} = -4\pi (e n_e v_e + e_i n_i v_i).$$

Let n_{e0} and n_{i0} be the values of the unperturbed electron and ion density (concentration) in a neutral plasma, respectively, so that $e_i n_{i0} + e n_{e0} = 0$. We determine $\omega_p = (4\pi e^2 n_{e0}/m_e)^{1/2}$ – the plasma frequency, $k_p = \omega_p/c$, and enter the dimensionless magnitudes

$$\rho = k_p x, \quad \theta = \omega_p t, \quad N_e = \frac{n_e}{n_{e0}}, \quad V_e = \frac{v_e}{c}, \quad P_e = \frac{p_e}{m_e c},$$

$$N_i = \frac{n_i}{n_{i0}}, \quad V_i = \frac{v_i}{c}, \quad E = -\frac{eE_x}{m_e c \omega_p}, \quad \delta = -\frac{m_e e_i}{m_i e} > 0.$$

In the new variables, system (5.1) will take the form

$$\frac{\partial N_e}{\partial \theta} + \frac{\partial}{\partial \rho}(N_e V_e) = 0, \quad \frac{\partial N_i}{\partial \theta} + \frac{\partial}{\partial \rho}(N_i V_i) = 0, \quad \frac{\partial E}{\partial \theta} = N_e V_e - N_i V_i,$$

$$\gamma = \sqrt{1 + P_e^2}, \quad V_e = \frac{P_e}{\gamma}, \quad \frac{\partial P_e}{\partial \theta} + E + \frac{\partial \gamma}{\partial \rho} = 0, \quad \frac{\partial V_i}{\partial \theta} + V_i \frac{\partial V_i}{\partial \rho} = \delta E. \tag{5.2}$$

From the first three equations (5.2) it follows that

$$\frac{\partial}{\partial \theta}\left(N_e - N_i + \frac{\partial E}{\partial \rho}\right) = 0.$$

This relationship is true both in the absence of plasma oscillations

($N_e \equiv N_i \equiv 1$, $E \equiv 0$), and in their presence, therefore, we have a simpler expression for the electron density:

$$N_e(\rho,\theta) = N_i(\rho,\theta) - \frac{\partial E(\rho,\theta)}{\partial \rho}. \tag{5.3}$$

Excluding the density N_e from system (5.2), we arrive at the P1EI equations, describing the free plane one-dimensional oscillating motions of electrons and ions in a cold ideal plasma:

$$\frac{\partial P_e}{\partial \theta} + E + \frac{\partial \gamma}{\partial \rho} = 0, \quad \gamma = \sqrt{1 + P_e^2}, \quad V_e = \frac{P_e}{\gamma},$$

$$\frac{\partial V_i}{\partial \theta} - \delta E + V_i \frac{\partial V_i}{\partial \rho} = 0, \quad \frac{\partial N_i}{\partial \theta} + N_i \frac{\partial V_i}{\partial \rho} + V_i \frac{\partial N_i}{\partial \rho} = 0, \tag{5.4}$$

$$\frac{\partial E}{\partial \theta} - N_i(V_e - V_i) + V_e \frac{\partial E}{\partial \rho} = 0.$$

Note that in addition to the desired variables, system (5.4) includes a small parameter δ, which essentially characterizes the mass ratio of the electron and the proton ($\delta \approx 1/1836$).

We discuss the boundary and initial conditions for the equations (5.4). If we assume that the oscillating part of the plasma occupies a limited part of the space (in this case, the segment $[-d, d]$) and all the perturbations do not go beyond its limits, then the boundary conditions

$$P_e(-d,\theta) = P_e(d,\theta) = 0, \quad V_i(-d,\theta) = V_i(d,\theta) = 0 \quad \forall \theta \geq 0 \tag{5.5}$$

fix the oscillating plasma slab, the boundaries of which do not depend on time. We clarify that here we are not talking about the spatial distribution of the density of charged particles in the form of a step function, which traditionally characterizes the plasma slab. Moreover, outside the oscillation region, the electron and ion densities are unperturbed, that is, they are equal to a constant that continuously extends the corresponding values from the interior of the region.

Additionally, we note that equations (5.4) of all solutions that do not depend on space and time can only be trivial, that is, $P_e = V_e \equiv V_i \equiv E \equiv 0$, $N_i \equiv 1$. Therefore, the basic system can be equipped with space-localized initial conditions for $\rho \in (-d, d)$

$$P_e(\rho,0) = P_{e_0}(\rho), V_i(\rho,0) = V_{i_0}(\rho),$$
$$N_i(\rho,0) = N_{i_0}(\rho), E(\rho,0) = E_0(\rho), \tag{5.6}$$

continuously adjacent to trivial solutions outside the plasma slab, i.e., $P_{e_0}(\pm d) = V_{i_0}(\pm d) = E_0(\pm d) = 0$, $N_{i_0}(\pm d) = 1$. Now from the locality of the initial conditions, the boundary condition for electric field disturbances are:

$$E(-d,\theta) = E(d,\theta) = 0 \quad \forall \theta \geq 0. \tag{5.7}$$

Summarizing all the above, the following initial–boundary problem can be defined for the plasma slab: find the semi-infinite strip $\{(\rho, \theta): \theta > 0, -d < \rho < d\}$ the solution of equations (5.4), satisfying the initial (5.6) and boundary (5.5), (5.7) conditions.

Let us choose the following functions as the initial conditions (5.6):

$$E_0(\rho) = \begin{cases} \left(\dfrac{a_*}{\rho_*}\right)^2 \rho \exp\left\{-2\dfrac{\rho^2}{\rho_*^2}\right\} & \text{at} \quad |\rho| < d, \\ 0 & \text{at} \quad |\rho| = d; \end{cases} \tag{5.8}$$

$$P_{e_0}(\rho) = 0, \quad V_{i_0}(\rho) = 0, \quad N_{i_0}(\rho) = 1 + \frac{\delta}{1+\delta} \frac{\partial E_0(\rho)}{\partial \rho}.$$

Such a perturbation of the electric field, which initiates oscillations, is characteristic of the passage of a short high-power laser pulse through the plasma, which has a Gaussian intensity distribution over space [45, 123]. By virtue of the exponential damping of the function $E_0(\rho)$, to ensure with sufficient accuracy the boundary condition (5.7), it suffices to set $d = 4.5\rho_\square$.

Given the symmetry of the problem under consideration, we note that if the initial functions $P_{e0}(\rho)$, $V_{i0}(\rho)$, $E_0(\rho)$ are odd functions, and $N_{i0}(\rho)$ is an even function relative to the origin, then, by virtue of the equations (5.4), the functions $P_e(\rho, \theta)$ (simultaneously with $V_e(\rho, \theta)$),

$V_i(\rho, \theta)$ and $E(\rho, \theta)$ are odd, and the function $N_i(\rho, \theta)$ is respectively even for all $\theta \geq 0$. This allows studies to be conducted not on the whole segment $[-d, d]$, but only on any half of it.

Recall that in the Eulerian variables the electron density function $N_e(\rho, \theta)$, which is defined by relation (5.3), is of fundamental importance for observing the oscillation breaking process.

5.2. Scaling equations and difference scheme

Given the presence of a small parameter δ, we scale the equations (5.4). We first do the replacement of the desired variables

$$V_i(\rho,\theta)= \delta W(\rho,\theta), \quad N_i(\rho,\theta)=1+\delta K(\rho,\theta), \tag{5.9}$$

and then we divide the equations for the new functions W and K by the quantity δ. Note that now the index ' $_e$ ', referring to the functions characterizing the dynamics of electrons, becomes uninformative and can be omitted without any loss of meaning. After these changes, system (5.4) takes the form

$$\frac{\partial P}{\partial \theta}+E+\frac{\partial \gamma}{\partial \rho}=0, \quad \gamma=\sqrt{1+P^2}, \quad V=\frac{P}{\gamma},$$

$$\frac{\partial W}{\partial \theta}-E+\delta W\frac{\partial W}{\partial \rho}=0, \quad \frac{\partial K}{\partial \theta}+(1+\delta K)\frac{\partial W}{\partial \rho}+\delta W\frac{\partial K}{\partial \rho}=0, \tag{5.10}$$

$$\frac{\partial E}{\partial \theta}-(1+\delta K)(V-\delta W)+V\frac{\partial E}{\partial \rho}=0.$$

The initial and boundary conditions for scaling will change only in the function $K(\rho, \theta)$, which describes the perturbation of the ion concentration:

$$K(\rho,\theta=0)=\frac{1}{1+\delta}\frac{\partial E_0(\rho)}{\partial \rho},0\leq \rho<d; \quad K(d,\theta)=0, \theta \geq 0. \tag{5.11}$$

Thus, we construct the difference scheme to find an approximate solution of equations (5.10) with initial and boundary conditions (5.5) – (5.8), (5.11).

Note that in system (5.10) the interaction of two physical processes is presented: nonlinear oscillations at a fixed point in space and their space–time transfer. Therefore, similarly to the previous chapters, as well as following the works [79, 94], we construct a splitting scheme for physical processes using the Lax–Wendroff ('tripod') scheme for the transfer equations [7].

We relate to the description of the process of nonlinear oscillations by the equation

$$\frac{\partial \tilde{P}}{\partial \theta} + \tilde{E} = 0, \quad \tilde{V} = \frac{\tilde{P}}{\sqrt{1 + \tilde{P}^2}}, \quad \frac{\partial \tilde{W}}{\partial \theta} - \tilde{E} = 0,$$

$$\frac{\partial \tilde{K}}{\partial \theta} + (1 + \delta \tilde{K}) \frac{\partial \tilde{W}}{\partial \rho} = 0, \quad \frac{\partial \tilde{E}}{\partial \theta} - (1 + \delta \tilde{K})(\tilde{V} - \delta \tilde{W}) = 0. \tag{5.12}$$

and to their transfer in space and time –

$$\frac{\partial \bar{P}}{\partial \theta} + \frac{\partial \bar{\gamma}}{\partial \rho} = 0, \quad \bar{\gamma} = \sqrt{1 + \bar{P}^2}, \quad \bar{V} = \frac{\bar{P}}{\bar{\gamma}},$$

$$\frac{\partial \bar{E}}{\partial \theta} + \bar{V} \frac{\partial \bar{E}}{\partial \rho} = 0, \quad \frac{\partial \bar{W}}{\partial \theta} + \delta \frac{\partial}{\partial \rho}\left(\frac{\bar{W}^2}{2}\right) = 0, \quad \frac{\partial \bar{K}}{\partial \theta} + \delta \bar{W} \frac{\partial \bar{K}}{\partial \rho} = 0. \tag{5.13}$$

As a basis for the time discretization of both systems, the usual leapfrog scheme is applicable [139]. Let τ be a time step, then we will assign to the 'integer' time moments $\theta_j = j\tau(j \geq 0 - $ integer) $E, \tilde{E}, \bar{E}, \tilde{K}, \bar{K}$, to 'half integer' $\theta_{j \pm 1/2} - P, \tilde{P}, \bar{P}, \tilde{W}, \bar{W}$, and also the values of γ and V depending on the momentum P. The choice of the corresponding time moment for the function value will be denoted by the superscript. For discretization in space, we will use a grid with a constant step h so that $\rho_m = mh$, $|m| \leq M$, $Mh = d$.

We write the difference equations approximating systems (5.12) and (5.13). For the first one we get

$$\frac{\tilde{P}_m^{j+1/2} - \tilde{P}_m^{j-1/2}}{\tau} + \tilde{E}_m^j = 0, \quad \tilde{V}_m^{j+1/2} = \tilde{P}_m^{j+1/2} / \sqrt{1 + \left(\tilde{P}_m^{j+1/2}\right)^2},$$

$$\frac{\tilde{W}_m^{j+1/2} - \tilde{W}_m^{j-1/2}}{\tau} - \tilde{E}_m^j = 0,$$

$$\frac{\tilde{K}_m^{j+1} - \tilde{K}_m^j}{\tau} + \left(1 + \delta \frac{\tilde{K}_m^{j+1} + \tilde{K}_m^j}{2}\right) \frac{\tilde{W}_{m+1}^{j+1/2} - \tilde{W}_{m-1}^{j+1/2}}{2h} = 0, \tag{5.14}$$

$$\frac{\tilde{E}_m^{j+1} - \tilde{E}_m^j}{\tau} - \left(1 + \delta \frac{\tilde{K}_m^{j+1} + \tilde{K}_m^j}{2}\right)(\tilde{V}_m^{j+1/2} - \delta \tilde{W}_m^{j+1/2}) = 0,$$

$$\tilde{P}_m^{j-1/2} = P_m^{j-1/2}, \tilde{E}_m^j = E_m^j, \tilde{W}_m^{j-1/2} = W_m^{j-1/2}, \tilde{K}_m^j = K_m^j, \quad |m| \leq M - 1.$$

Before writing down the approximation of system (5.13), we recall that the time discretization in the 'tripod' scheme for a model equation (such as nonlinear transfer)

$$\frac{\partial u}{\partial t} + \frac{\partial G(u)}{\partial x} = 0, \quad G(u) = \frac{u^2}{2}$$

has the appearance

$$\frac{u^{j+1} - u^j}{\tau} + \frac{\partial G^j}{\partial x} = \frac{\tau}{2} \frac{\partial}{\partial x}\left(A^j \frac{\partial G^j}{\partial x}\right),$$

where $A = \dfrac{\partial G}{\partial u}$, and the superscript denotes the belonging of the function to the corresponding point in time. If, however, as the model equation we take the linear transport equation

$$\frac{\partial u}{\partial t} + v \frac{\partial u}{\partial x} = 0,$$

then the time discretization, analogous to the 'tripod' scheme, for it, respectively, looks like this:

$$\frac{u^{j+1} - u^j}{\tau} + \left(v^j + \frac{\tau}{2}\frac{\partial v}{\partial t}\right)\frac{\partial u}{\partial x} = \frac{\tau v^j}{2}\frac{\partial}{\partial x}\left(v^j \frac{\partial u}{\partial x}\right),$$

and also has an $O(\tau^2)$ approximation on smooth solutions.

The discrete analogue of system (2.5), convenient for realization, built on the basis of the given model schemes, has the following form:

$$\frac{\overline{P}_m^{j+1/2} - \overline{P}_m^{j-1/2}}{\tau} + \overline{\gamma}_{\bar{X},m}^{j-1/2} = \frac{\tau}{2}\left(\overline{V}_{s,m}^{j-1/2}\overline{\gamma}_{x,m}^{j-1/2}\right)_{\bar{X},m},$$

$$\overline{\gamma}_m^{j+1/2} = \sqrt{1 + \left(\overline{P}_m^{j+1/2}\right)^2}, \quad \overline{V}_m^{j+1/2} = \frac{\overline{P}_m^{j+1/2}}{\overline{\gamma}_m^{j+1/2}},$$

$$\frac{\overline{E}_m^{j+1} - \overline{E}_m^j}{\tau} + \left(\overline{V}_m^{j+1/2} + \frac{\tau}{2}\frac{\overline{V}_m^{j+1/2} - \overline{V}_m^{j-1/2}}{\tau}\right)\overline{E}_{\bar{X},m}^j =$$

$$= \frac{\tau}{2}\overline{V}_m^{j+1/2}\left(\overline{V}_{s,m}^{j+1/2}\overline{E}_{x,m}^j\right)_{\bar{X},m},$$

$$\frac{\overline{W}_m^{j+1/2} - \overline{W}_m^{j-1/2}}{\tau} + \overline{G}_{\ddot{X},m}^{j-1/2} = \frac{\delta\tau}{2}\left(\overline{W}_{s,m}^{j-1/2}\overline{G}_{x,m}^{j-1/2}\right)_{\overline{X},m},$$

$$\frac{\overline{K}_m^{j+1} - \overline{K}_m^{j}}{\tau} + \delta\left(\overline{W}_m^{j+1/2} + \frac{\tau}{2}\frac{\overline{W}_m^{j+1/2} - \overline{W}_m^{j-1/2}}{\tau}\right)\overline{K}_{\ddot{X},m}^{j} =$$

$$= \frac{\delta^2\tau}{2}\overline{W}_m^{j+1/2}\left(\overline{W}_{s,m}^{j+1/2}\overline{K}_{x,m}^{j}\right)_{\overline{X},m},$$

$$\overline{P}_m^{j-1/2} = \tilde{P}_m^{j+1/2}, \overline{E}_m^{j} = \tilde{E}_m^{j+1}, \overline{W}_m^{j-1/2} = \tilde{W}_m^{j+1/2}, \overline{K}_m^{j} = \tilde{K}_m^{j+1}, \quad |m| \le M-1,$$

$$\overline{P}_0^{j+1/2} = \overline{P}_M^{j+1/2} = \overline{E}_0^{j+1} = \overline{E}_M^{j+1} = \overline{W}_0^{j+1/2} = \overline{W}_M^{j+1/2} = 0.$$

(5.15)

In expression (5.15) the following notation is used:

$$\overline{G}^{j-1/2} \equiv \overline{G}\left(\overline{W}^{j-1/2}\right) = \frac{\delta}{2}\left(\overline{W}^{j-1/2}\right)^2, F_{\ddot{X},m} = (F_{m+1} - F_{m-1})/(2h)$$

– the central difference,

$$F_{X,m} = (F_{m+1} - F_m)/h \text{ and } F_{\overline{X},m} = (F_m - F_{m-1})/h$$

– differences forward and backward respectively, $F_{s,m} = (F_{m+1} + F_m)/2$.

After calculations using the scheme (5.15), we should redefine the required functions at the following time layer:

$$P_m^{j+1/2} = \overline{P}_m^{j+1/2}, E_m^{j+1} = \overline{E}_m^{j+1}, W_m^{j+1/2} = \overline{W}_m^{j+1/2}, K_m^{j+1} = \overline{K}_m^{j+1}, \quad |m| \le M,$$

and calculate (if necessary) the electron density perturbation ($N_e(\rho, \theta) = 1 + N(\rho, \theta)$) using the formula

$$N_m^{j+1} = \begin{cases} \delta K_m^{j+1} - \dfrac{E_{m+1}^{j+1} - E_{m-1}^{j+1}}{2h} & \text{at} \quad |m| \le M-1, \\ 0 & \text{at} \quad |m| = M. \end{cases}$$

(5.16)

At this time, the calculations at the *j*-th time step end and we can proceed to the next step. It should be noted that the initial data (5.8) correspond to $\theta = 0$; therefore, they should be attributed for P and W to the layer with number $-1/2$, and for E and K – to number 0.

We make comments on the scheme under consideration (5.14), (5.15). For each of the auxiliary problems with sufficient smoothness of the solution there is an approximation of the order $O(\tau^2 + h^2)$, which for the total splitting scheme [41] leads to the order $O(\tau + h^2)$. In addition, scheme (5.14) is unconditionally stable, and for scheme (5.15), the condition of stability of the form $\tau = O(h)$, obtained on the basis of the spectral attribute [7, 20], is valid. The latter makes it possible to achieve substantial savings in computational resources due to a weaker stability condition without loss of approximation. In addition, the scheme (5.14), (5.15) can be implemented using explicit formulas, which gives rise to the potential for parallelization when generalized to multidimensional cases. Note also that in the case of stationary ions ($\delta = 0$) the scheme was presented and tested in [94], where the results of calculations of relativistic oscillations in the Eulerian variables were indistinguishable from the results of calculations in the Lagrangian variables up to breaking.

5.3. Axial solution

In [100] (see also section 2.2), the concept of an axial solution was introduced for nonlinear problems describing laser-plasma interactions and possessing axial symmetry as a solution having a local-linear dependence on the spatial coordinate.

In this section, the axial solution is used to establish the possibility of a new type of breaking long-lived electron oscillations in a plasma. We specify that the axial solution of equations (5.10) is understood as a real solution of the form

$$V(\rho,\theta)=V_a(\theta)\rho, \quad W(\rho,\theta)=W_a(\theta)\rho, \quad E(\rho,\theta)=E_a(\theta)\rho,$$
$$K(\rho,\theta)=K_a(\theta), \quad \gamma(0,\theta)=\gamma_a(\theta)\equiv 1.$$

It is easy to make sure that the time-dependent factors in this case satisfy the system of ordinary differential equations

$$V_a'=-(E_a+V_a^2), \quad E_a'=(1+\delta K_a)(V_a-\delta W_a)-V_aE_a,$$
$$W_a'=E_a-\delta W_a^2, \quad K_a'=-(1+\delta K_a)W_a. \tag{5.17}$$

We supplement the equations obtained by the initial conditions following from (5.8), (5.11)

$$V_a(0)=W_a(0)=0, \quad E_a(0)=\alpha, \quad K_a(0)=\frac{\alpha}{1+\delta}, \tag{5.18}$$

where $\alpha = (a_*/\rho_*)^2$. If necessary, the corresponding perturbations of the electron density on the axis, i.e., $N_a(\theta)$, can be calculated from the formula

$$N_a(\theta) = \delta K_a(\theta) - E_a(\theta).$$

Note that the reduced Cauchy problem (5.17), (5.18) is not trivial, since even in the particular case $\delta = 0$ admits both regular periodic solutions, discussed in section 2.2 (see also [102]) and solutions that have singularities in the first period of oscillations (the so-called blow-up solutions [157]).

We will conduct an asymptotic analysis of long-lived solutions of problem (5.17), (5.18) with respect to the expansion of the desired functions in a small parameter δ [25]:

$$F(\theta) = F_0(\theta) + \delta F_1(\theta) + \delta^2 F_2(\theta) + \ldots.$$

Zero approximation satisfies the equations

$$V_0' = -(E_0 + V_0^2), \quad E_0' = V_0 - V_0 E_0, \quad W_0' = E_0, \quad K_0' = -W_0 \qquad (5.19)$$

and initial conditions

$$V_0(0) = W_0(0) = 0, \quad E_0(0) = K_0(0) = \alpha. \qquad (5.20)$$

For the first two functions, for $\alpha < 1/2$ of the formulas in section 2.2, in the particular case of $\beta = 0$ the analytical expressions follow

$$V_0(\theta) = -\frac{s \sin\theta}{1 + s\cos\theta}, \quad E_0(\theta) = \frac{s\cos\theta}{1 + s\cos\theta}, \quad \text{where} \quad s = \frac{\alpha}{1-\alpha} < 1. \qquad (5.21)$$

This makes it possible to construct in an explicit form the function $W_0(\theta)$. First, we calculate the period average of the function $E_0(\theta)$:

$$I = \int_0^{2\pi} E_0(\theta)\,d\theta = \int_0^{2\pi} \frac{s\cos\theta}{1 + s\cos\theta}\,d\theta = 2\pi\left(1 - \frac{1}{\sqrt{1-s^2}}\right) < 0, \qquad (5.22)$$

and then, using (5.22), we write

$$W_0(\theta) = Ik + \int_0^{\theta'} \frac{s\cos x}{1+s\cos x}dx,$$

$$\text{where} \quad \theta = 2\pi k + \theta', k \geq 0, \ 0 \leq \theta' < 2\pi. \tag{5.23}$$

The representation of the argument $\theta \bmod 2\pi$ is related to the well-known formula [51]

$$\int \frac{s\cos x}{1+s\cos x}dx = x - \frac{2}{\sqrt{1-s^2}}\arctan\frac{(1-s)\tan\frac{x}{2}}{\sqrt{1-s^2}}, \quad |s|<1.$$

It follows from formulas (5.22), (5.23) that the radial derivative of the velocity of ions on the axis of symmetry, that is, the function $W_0(\theta)$, is always negative. In other words, in the vicinity of the axis, the velocity of the ions is always directed toward the axis, which means that the concentration of ions in the vicinity of the axis monotonously (from period to period) increases. In turn, this means that the amplitude of the electron oscillations must also increase, since their acceleration in the vicinity of the axis of symmetry is directly related to the concentration of ions on the axis. Thus, the starting oscillation amplitude on the axis of symmetry, equal to α, should sooner or later exceed the critical value $1/2$ (see section 2.2 for a comment on Theorem 2.2.1, and also [102]), which is guaranteed to lead to the 'blow-up' effect for the next period of oscillations.

We give two illustrations to the above. Let us apply the classical Runge–Kutta formula of the fourth order of accuracy [20] for the numerical solution of the Cauchy problem (5.17), (5.18) in the normal form $\mathbf{y}' = \mathbf{f}(t, \mathbf{y})$ with the given initial condition $\mathbf{y}(0)$:

$$\mathbf{y}_{j+1} = \mathbf{y}_j + \frac{1}{6}(\mathbf{k}_1 + 2\mathbf{k}_2 + 2\mathbf{k}_3 + \mathbf{k}_4),$$

where

$$\mathbf{k}_1 = \tau\mathbf{f}(t,\mathbf{y}), \quad \mathbf{k}_2 = \tau\mathbf{f}(t+\tau/2, \mathbf{y}+\mathbf{k}_1/2),$$
$$\mathbf{k}_3 = \tau\mathbf{f}(t+\tau/2, \mathbf{y}+\mathbf{k}_2/2), \quad \mathbf{k}_4 = \tau\mathbf{f}(t+\tau, \mathbf{y}+\mathbf{k}_3).$$

Figure 5.1 shows the perturbations of the electron density $N_a(\theta)$ calculated for $\alpha = 0.3$ with a step $\tau = 1/640$. The dotted line

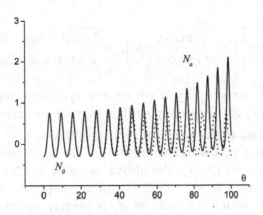

Fig. 5.1. Perturbation of electron density on the axis of symmetry: N_a – ions are mobile (solid line), N_0 – ions are fixed (dashed line).

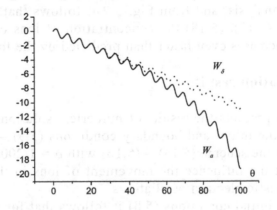

Fig. 5.2. Ion velocity on the axis of symmetry: W_a – numerical solution (solid line), W_δ – asymptotic solution (dashed line).

denotes the time dependence of the density in the case of stationary ions ($\delta = 0$), that is, $N_0(\theta) = -E_0(\theta)$, therefore the calculation results coincide with formula (5.21). Here it is easy to trace the 2π-periodicity of limited oscillations over the entire time interval under consideration. The solid line in fig. 5.1 characterizes the electron density perturbation when ions are mobile ($\delta = 1/2000$). In turn, a monotonous increase in amplitude («swinging») is well observed here, which leads to breaking of oscillations.

Figure 5.2 shows for the same parameters α and τ for mobile ions ($\delta = 1/2000$) a comparison of the zero approximation $W_\delta(\theta)$:

$$W_\delta(\theta) = \frac{1}{\sqrt{1+\delta}}\left[Ik + \int_0^{\theta'} \frac{s\cos x}{1+s\cos x}dx \right], \quad \begin{array}{c} \sqrt{1+\delta}\theta = 2\pi k + \theta', \\ k \geq 0, 0 \leq \theta' < 2\pi \end{array}, \quad (5.24)$$

with the scaled ion velocity $W_a(\theta)$ on the symmetry axis. Note that in formula (5.24) the frequency shift is taken into account: to match the linear oscillation frequency in (5.17), where $\omega = \sqrt{1+\delta}$, and also to avoid resonant (growing) solutions in the asymptotic expansion [25]. In the formula (5.24), the above ω value is chosen, although the value close to it $\omega \approx 1 + \delta/2$ is quite acceptable. Note that the frequency shift, which depends on δ, is clearly visible in Fig. 5.1, where the period of electronic oscillations, taking into account the dynamics of ions, is reduced compared with 2π.

Thus, numerical experiments are in good agreement with asymptotic analysis, and from Fig. 5.2 it follows that in the model under study (5.17), (5.18) the concentration of ions on the axis of symmetry increases even faster than predicted by the formula (5.24).

5.4. Calculation results

This section presents the results of numerical solution of equations (5.10) with the initial and boundary conditions (5.5) – (5.8), (5.11) according to the scheme (5.14) – (5.16) with $\delta = 1/2000$, i.e., taking into account the influence the movement of ions in the process of plane relativistic electron oscillations.

From the initial conditions (5.8) it follows that for a fixed value of δ the oscillations depend on two parameters, a_* and ρ_*, or, which is the same, on $\alpha = (a_*/\rho_*)^2$ and ρ_*. In chapter 3, devoted exclusively to relativistic electron oscillations (see also work [94]), it was found that for any fixed $\alpha < 1/2$, the breaking time depends on the parameter ρ_*. Moreover, with a decrease in the parameter ρ_*, the time of breaking increases, and, accordingly, with increasing – decreases. The pattern of breaking is always the same, namely, off-axis. On the other hand, in the study in the previous section of the axial solution of system (5.10), which is determined by only one parameter α, it was formulated an assumption about the admissibility of another type of breaking oscillation – axial, which occurs due to the directional movement of ions in the vicinity of the axis of symmetry of the region. Moreover, a monotonous increase in α (but without passing

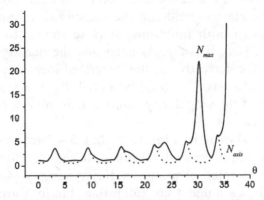

Fig. 5.3. Dynamics of electron density in off-axis breaking: N_{max} – maximum in the region (solid line), N_{axis} – value at $\rho = 0$ (dashed line).

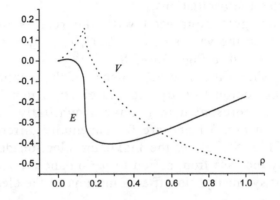

Fig. 5.4. Spatial distributions of odd functions: electric field E and electron velocity V at the moment of off-axis breaking; E – solid line, V – dashed line.

through the critical value 1/2) should lead to a decrease in the time of axial breaking.

We first consider the results of numerical experiments with the values of $a_* = 1.95$, $\rho_* = 5$, and $\tau = h = 1/1500$, which are presented in Figs. 5.3–5.5.

Figure 5.3 shows the time dependence of the maximum electron density in the region $[-d, d]$ and the electron density at $\rho = 0$. It is easy to see that in the initial oscillation periods the maximum density values are concentrated on the symmetry axis, then (in the fourth period) a local off-axis maximum, the growth of which further leads to the breaking of oscillations. The amplitudes of the axial oscillations in this case also grow, but much slower.

Figure 5.4 shows the dependence on the spatial coordinates of the functions of the electric field and the velocity of electrons. By virtue of the oddness of both functions, it is sufficient to observe them only for $\rho \geq 0$. Here it is clearly seen that the density singularity is determined by the derivative of the generated local-step function $E(\rho)$ and is located off-axis. It should be noted that the electron velocity in the vicinity of the singularity point is continuous, but has a jump in the derivative.

The graphs shown in Figs. 5.3 and 5.4, are typical for off-axis breaking, regardless of taking into account the dynamics of ions, considering relativism and spatial symmetry [46, 76, 79, 94]. However, in the case under consideration, the ions are mobile, their spatial distribution differs from the constant (see Fig. 5.5), but the cause of the breaking in the described calculations is 'independent' relativistic electron oscillations.

Let us now get acquainted with the results of numerical experiments with the values of $a_* = 2.07$, $\rho_* = 3$, $\tau = h = 1/1500$, which are presented in Figs. 5.6–5.8.

Figure 5.6 shows the time dependence of the maximum electron density in the region $[-d, d]$ and the electron density at $\rho = 0$. The functions represented in it have a qualitative similarity to the functions in Fig. 5.1 and are fundamentally different from the functions in Fig. 5.3. Here, the maximum electron density values monotonously increase from period to period and are located strictly on the axis of symmetry. No off-axis maxima of the electron density

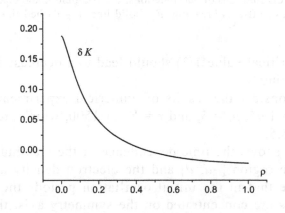

Fig. 5.5. Spatial distribution at $\rho \geq 0$ of ion density perturbations δK (even function) at the moment of off-axis breaking.

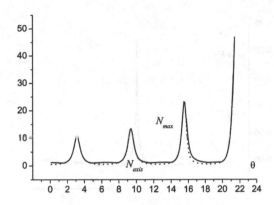

Fig. 5.6. Dynamics of electron density in axial breaking: N_{max} – maximum in the region (solid line), N_{axis} – value at $\rho = 0$ (dashed line).

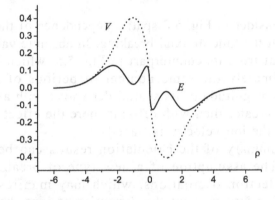

Fig. 5.7. The spatial distributions of the electric field E and the electron velocity V at the time of axial tipping: E – solid line, V – dotted line.

function are observed until breaking. Quantitative differences from Fig. 5.1 are associated with the difference in the value of the parameter α: in this case, it is equal to about 0.476, which led to rapid axial breaking over the fourth period of oscillation.

Figure 5.7 shows the dependences on the spatial coordinate of the functions of the electric field and the velocity of electrons upon axial breaking. As in Fig. 5.4, we observe that the singularity of the electron density is determined by the derivative of the generated locally-step function $E(\rho)$, however, a jump in its values is located here at $\rho = 0$. Pay attention to the spatial dependence of the electron velocity in axial breaking: similar to the step, i.e., discontinuous function. Such a structure is qualitatively different from the velocity function shown in Fig. 5.4.

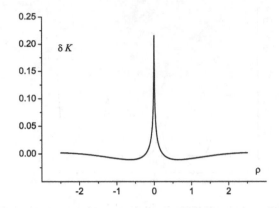

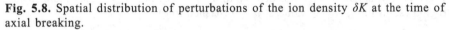

Fig. 5.8. Spatial distribution of perturbations of the ion density δK at the time of axial breaking.

Finally, consider in Fig. 5.8 spatial dependence of the ion density perturbation at the time of axial breaking. In absolute values, it is not much different from its counterpart in Fig. 5.5, which is associated with approximately the same number of periods of its increase. However, the magnitude of its spatial derivative with axial breaking is noticeably greater than with off-axis: here the effect of the radial derivative of the ion velocity is traced.

A brief summary of the calculation results can be formulated as follows. The assumption of a new type of breaking of plane relativistic electron oscillations, which may manifest itself as a result of taking into account the dynamics of ions, has been fully confirmed. The parameters of the initial problem are determined, for which the difference in breaking types (off-axis and axial) is very intuitive. However, it should be noted that so far it has not been possible to select such task parameters so that both types of breaking can be realized simultaneously. It is also unknown whether this is possible in principle.

5.5. Bibliography and comments

The chapter considers highly nonlinear one-dimensional plane plasma oscillations. Such oscillations can occur in a wake wave excited by a short laser pulse in a rarefied plasma when it is focused by a cylindrical lens. With this method of focusing, the focal spot of a laser pulse is a strip, the length of which far exceeds its transverse size. Therefore, in this case, electrons, under the action of ponderomotive forces of laser radiation, will actually perform one-dimensional plane oscillations.

A preliminary perturbation method has been used to theoretically study the influence of ion dynamics on the relativistic breaking of electron oscillations. Analysis of the zero asymptotic approximation allowed us to formulate the assumption that there is another type of breaking of long-lived electronic oscillations that is different from the previously known one. Then, using numerical simulation methods, it was shown that, depending on the initial plasma parameters, two breaking scenarios can be realized in the plane relativistic case. Both of them, for solving the corresponding mathematical problem, lead to the so-called gradient catastrophe, when the solution is limited, and the spatial derivative has a singularity [81, 121]. Their distinction is particularly evident with some specific symmetry of the initial conditions, in particular, when the electric field exciting the oscillations is odd and linear in the vicinity of the axis of symmetry of the region occupied by the plasma. In this case, the breaking can be realized either on the axis of symmetry (a new scenario for long-lived oscillations), or outside it (a scenario previously known).

It should be noted that axial breaking was considered earlier in [119]. However, the conclusion about its capabilities was made on the basis of oscillations of electrons only. The principal difference between the results of the chapter and the conclusions made in [119] is that the breaking on the axis of symmetry arises solely as a result of taking into account the motion of ions. In the process of oscillation of electrons over time, as a result of the action of an electric field of charge separation, ions move slowly to the axis of symmetry, as a result of which their concentration there increases markedly. It is this effect that leads to axial breaking. Without taking into account the movement of ions, the breaking of plasma oscillations and wake waves, in accordance with the conclusions of this book, occurs strictly outside the axis of the laser pulse. Of course, it should be clarified that we are talking about fluctuations that continue to go on for more than half of one period.

Numerical experiments were conducted using a specially designed difference scheme in the Eulerian variables based on splitting along physical processes. In the case of fixed ions, the scheme was presented and tested in [94], where the results of calculations in the Eulerian variables were indistinguishable from the results of calculations in the Lagrangian variables up to breaking. A part of the calculations concerning the accuracy control of the constructed algorithm was carried out on the SKIF MSU Chebyshev supercomputer.

It can be concluded that, for certain plasma parameters, taking into account the movement of ions qualitatively changes the dynamics of electron oscillations, which can significantly affect the interpretation of the results of field experiments.

The results of the chapter were obtained in [95].

6

Plane two-dimensional relativistic electron oscillations

The chapter deals with a numerical–asymptotic model of breaking plane two-dimensional relativistic electron oscillations. The model completely reproduces one-dimensional cylindrical oscillations, but the main interest is the study of deviations from axial symmetry. The asymptotic theory uses the construction of uniformly suitable in time solutions of weakly non-linear equations. Numerical simulation is carried out using a special algorithm built on the basis of the finite-difference method and using shifted grids.

6.1. Formulation of the problem

We will consider, as before, the plasma to be a cold, ideal, relativistic electron liquid, neglecting collisional and recombination effects, as well as the movement of ions. Then, based on the results of section 1.2 the system of equations (1.5)–(1.12) can be represented in the following form:

$$\frac{\partial n}{\partial t} + \operatorname{div}(n\mathbf{v}) = 0, \quad \frac{\partial \mathbf{p}}{\partial t} = e\mathbf{E} - mc^2 \nabla \gamma, \quad \frac{\partial \mathbf{B}}{\partial t} = -\operatorname{curl} \mathbf{E},$$

$$\gamma = \sqrt{1 + \frac{|\mathbf{p}|^2}{m^2 c^2}}, \quad \mathbf{v} = \frac{\mathbf{p}}{m\gamma}, \quad \frac{\partial \mathbf{E}}{\partial t} = -4\pi e n\mathbf{v} + \operatorname{curl} \mathbf{B},$$

$$(6.1)$$

where e, m is the electron charge and mass ($e < 0$), c is the speed of light, n, $\mathbf{p}$, $\mathbf{v}$ is the concentration (density), momentum and velocity of electrons, γ is the Lorentz factor, $\mathbf{E}$, $\mathbf{B}$ are the vectors of electric and magnetic fields.

Consider the basic equations (6.1) under the assumption that the solution is determined only by the x- and y- components of the vector functions $\mathbf{p}$, $\mathbf{v}$, $\mathbf{E}$, and z-the component of the vector function $\mathbf{B}$ and there is no dependence on the variable z. Then two-dimensional scalar equations can be written in a dimensionless form:

$$\frac{\partial P_x}{\partial \theta} + E_x + \frac{1}{2\gamma}\frac{\partial}{\partial x}\left(P_x^2 + P_y^2\right) = 0,$$

$$\frac{\partial P_y}{\partial \theta} + E_y + \frac{1}{2\gamma}\frac{\partial}{\partial y}\left(P_x^2 + P_y^2\right) = 0,$$

$$\gamma = \sqrt{1 + P_x^2 + P_y^2}, \quad \frac{\partial E_x}{\partial \theta} = N\frac{P_x}{\gamma} + \frac{\partial B_z}{\partial y}, \quad \frac{\partial E_y}{\partial \theta} = N\frac{P_y}{\gamma} - \frac{\partial B_z}{\partial x},$$

$$N = 1 - \left(\frac{\partial E_x}{\partial x} + \frac{\partial E_y}{\partial y}\right), \quad \frac{\partial B_z}{\partial \theta} = -\left(\frac{\partial E_y}{\partial x} - \frac{\partial E_x}{\partial y}\right).$$

(6.2)

The formula for the electron density N is a consequence of the first and last equations (6.1).

The basis for the non-dimensionalization of the variables, in addition to the above constants, is: n_0 is the value of the unperturbed electron density and $\omega_p = (4\pi e^2 n_0/m)^{1/2}$ is the plasma frequency. In particular, the following notation for the variables is used:

$$\theta = \omega_p t, \quad x = \frac{\omega_p}{c}x_1, \quad y = \frac{\omega_p}{c}x_2, \quad N = \frac{n}{n_0}.$$

The dimensionless components of the electric field (E_x, E_y), momentum (P_x, P_y) and magnetic field B_z are denoted by the same symbols in order to preserve the semantic continuity with the original equations.

The initial conditions at $\theta = 0$ are defined as

$$E_x = \left(\frac{a_*}{\rho_*}\right)^2 x\exp^2\left\{-\frac{x^2 + \alpha y^2}{\rho_*^2}\right\}, \quad P_x = P_y = 0,$$

$$E_y = \left(\frac{a_*}{\rho_*}\right)^2 \alpha y\exp^2\left\{-\frac{x^2 + \alpha y^2}{\rho_*^2}\right\}, \quad B_z = 0,$$

(6.3)

where the parameter α defines the symmetry properties of the problem under consideration. For $\alpha = 1$, the solution has axial symmetry, that is, all the desired functions depend only on the quantities $\rho = \sqrt{x^2 + y^2}$ and θ. In particular, this simulates the situation when a wake wave is initiated by a short laser pulse having a circular cross section. When $\alpha = 1 + \Delta$, $0 < |\Delta| < 1$, the number of symmetry axes of the solution becomes smaller: no more than two (including parity/oddness), which corresponds to the shape of the ellipse for the cross section of the initiating laser pulse. Note that the initial electric field in (6.3) is irrotational, that is, it can be represented as a gradient of a function that simulates the spatial distribution of the intensity of a laser pulse.

6.2. Asymptotic theory

Consider nonlinear plasma oscillations that at the time $\theta = 0$ are started by the initial conditions (6.3), and for convenience, we introduce

$$E_x(x, y, \theta = 0) = A(x, y), \quad E_y(x, y, \theta = 0) = B(x, y). \tag{6.4}$$

In this case, we assume that the initial electric field has an insignificant value in any metric: $\|A\| \ll 1$, $\|B\| \ll 1$. It is convenient to keep in mind the uniform metric for continuous functions, i.e. $\|f\| = \max_{x,y} |f(x, y)|$. Recall that if the parameter α in (6.3) is equal to unity, then under the action of an arbitrarily weak electric field axially symmetric plasma oscillations swing, which results in their breaking. Below, we will consider plasma oscillations for an arbitrary value of α based on the trajectories of electrons, which in dimensionless variables are of the form:

$$x(\theta) = x_0 + R_x(x_0, y_0, \theta), \quad y(\theta) = y_0 + R_y(x_0, y_0, \theta),$$
$$\frac{\partial R_x}{\partial \theta} = V_x, \quad \frac{\partial R_y}{\partial \theta} = V_y, \quad \frac{\partial P_x}{\partial \theta} = -E_x, \quad \frac{\partial P_y}{\partial \theta} = -E_y, \tag{6.5}$$

where (x_0, y_0) are the initial coordinates of the electron, (R_x, R_y) are small displacements of the electron from the initial position, and the components of the momentum $\mathbf{P}$ and the velocity $\mathbf{V}$ of the electron are related by

$$P = \frac{V}{\sqrt{1-|V|^2}} \approx V\left(1 + \frac{V_x^2 + V_y^2}{2}\right). \tag{6.6}$$

We will consider the situation when the region of localization of oscillations has sufficiently large dimensions ($\rho_* \gg 1$), which allows us to neglect the magnetic field outside a small neighbourhood of the origin. In this case, the Maxwell equations for the components of the electric field of plasma oscillations **E** have the form

$$\frac{\partial \mathbf{E}}{\partial \theta} = N\mathbf{V}, \quad N = 1 - \left(\frac{\partial E_x}{\partial x} + \frac{\partial E_y}{\partial y}\right), \tag{6.7}$$

where N is the dimensionless density of electrons.

Using relations (6.5)–(6.7), we obtain the following equations for the electron velocity components:

$$\left(\frac{\partial^2}{\partial \theta^2} + 1\right)V_x + \frac{1}{2}\frac{\partial^2}{\partial \theta^2}V_x(V_x^2 + V_y^2) = 0,$$
$$\left(\frac{\partial^2}{\partial \theta^2} + 1\right)V_y + \frac{1}{2}\frac{\partial^2}{\partial \theta^2}V_y(V_x^2 + V_y^2) = 0, \tag{6.8}$$

which are valid under the condition $\rho_* \gg 1$. Solving the system of equations (6.8) according to the perturbation theory [25, 154] with the initial conditions (6.3), (6.4), we obtain the following expressions for the trajectories of electrons

$$x = x_0 + A(x_0, y_0)\cos(\omega\theta), \quad y = y_0 + B(x_0, y_0)\cos(\omega\theta), \tag{6.9}$$

where the frequency ω of weakly non-linear plasma oscillations, taking into account the relativistic dependence of the electron mass on velocity, has the form

$$\omega = 1 + \Delta\omega, \quad \Delta\omega = -\frac{3}{16}(A^2 + B^2). \tag{6.10}$$

Recall that the shift $\Delta\omega$ of the fundamental oscillation frequency, as usual, arises from the condition of eliminating resonant terms, leading to time-increasing solutions.

Consider the change in the electron density function as a result of particle displacement from the initial position: it is determined by the formula

$$N = \left[\frac{\partial(x,y)}{\partial(x_0,y_0)} \right]^{-1}. \tag{6.11}$$

Calculating the Jacobian in the denominator of the formula (6.11), we find the dependence of the density on time:

$$N = \frac{1}{1 + G(x_0,y_0)\cos(\omega\theta) + \theta\sin(\omega\theta)F(x_0,y_0)}, \tag{6.12}$$

where

$$G(x_0,y_0) = \frac{\partial A}{\partial x_0} + \frac{\partial B}{\partial y_0}, \quad F(x_0,y_0) = -\left[A\frac{\partial\omega}{\partial x_0} + B\frac{\partial\omega}{\partial y_0} \right]. \tag{6.13}$$

From relation (6.12), it follows that over time, the density increases, and at some point in time its singularity may occur even at small values of $\|A\|$ and $\|B\|$.

Let us analyze in more detail the spatial dependence of functions (6.13). Moving from the Cartesian to polar coordinates, that is, assuming that $\rho_0 = \sqrt{x_0^2 + y_0^2}$, $\varphi = \arctan(y_0/x_0)$, after simple calculations we get the following expressions for the functions G and F:

$$G(\rho,\varphi) = \left(\frac{a_*}{\rho_*}\right)^2 \exp\left\{-2\rho^2\left(\cos^2\varphi + \alpha\sin^2\varphi\right)\right\} \times$$

$$\times\left\{1 + \alpha - 4\rho^2\left(\cos^2\varphi + \alpha^2\sin^2\varphi\right)\right\},$$

$$\tag{6.14}$$

$$F(\rho,\varphi) = \frac{3}{8}\left(\frac{a_*}{\rho_*}\right)^6 \rho_*^2\exp\left\{-6\rho^2\left(\cos^2\varphi + \alpha\sin^2\varphi\right)\right\} \times$$

$$\times\rho^2\left\{\cos^2\varphi + \alpha^3\sin^2\varphi - 4\rho^2\left(\cos^2\varphi + \alpha^2\sin^2\varphi\right)^2\right\}.$$

Here, for convenience, the notation $\rho = \rho_0/\rho_*$ is used. Since the value $|F(\rho, \varphi)| \ll 1$ by assuming that A and B are small, then, in accordance with formula (6.12), the density become infinite at time $\theta_{wb} \approx |F|^{-1}$. At the same time, the coversion time for the density to infinity corresponds to the absolute maximum of the function $|F|$.

We first consider the case of a small deviation from axial symmetry, when the parameter α is slightly different from unity: $\alpha = 1 + \Delta$, $|\Delta| \ll 1$. In this case, it is convenient to rewrite formula (6.14) as follows:

$$F(\rho,\varphi) = \frac{3}{8}\left(\frac{a_*}{\rho_*}\right)^6 \rho_*^2 \exp\{-6\rho^2\} \times$$
$$\times \rho^2 \left\{1 - 4\rho^2 + \left(3 - 22\rho^2 + 24\rho^4\right)\Delta\sin^2\varphi\right\}. \qquad (6.15)$$

When $\Delta = 0$, the modulus of the function in (6.15) has an absolute maximum at $\rho = 1/\sqrt{12}$, which fully corresponds to the axial symmetry of the process of destruction of electronic oscillations. In the vicinity of this value, function (6.15) has the form

$$F(\rho,\varphi) = \frac{1}{48}\left(\frac{a_*}{\rho_*}\right)^6 \frac{\rho_*^2}{\sqrt{e}}\left\{1 + 2\Delta\sin^2\varphi\right\}. \qquad (6.16)$$

From formula (6.16), it is easy to see that for $\Delta > 0$, which corresponds to $\alpha > 1$, the function $|F(\rho, \varphi)|$ has maxima at two points $\varphi = \pi/2$ and $\varphi = 3\pi/2$, where, in accordance with the formula $\theta_{wb} \approx |F|^{-1}$, the density grows to infinity. For a small negative value of Δ (that is, for $\alpha < 1$), the absolute value of the function from (6.16) is maximum when the polar angle $\varphi = 0$ and $\varphi = \pi$. Therefore, at these points at a distance of $\rho_0 = \rho_* / \sqrt{12}$, in this case, the singularity of the electron density arises.

We now consider the case of an arbitrary positive value of the parameter α. A numerical and analytical analysis of the dependence of the function $F(\rho, \varphi)$ on the coordinates in (6.14) allows us to draw the following conclusions. For $\alpha > 1$, the maximum absolute value of the function in (6.14) $F_{\max}(\rho,\varphi) = (3/16)(a_* / \rho_*)^6 \alpha^2 \rho_*^2 / (9\sqrt{e})$ is achieved at $\varphi = \pi/2$ and $\varphi = 3\pi/2$ at a distance from the axis $\rho_0 = \rho_* / (2\sqrt{3}\alpha)$. Hence we conclude that the time of breaking when $\alpha > 1$ is determined by the formula

$$\theta_{wb} \approx \frac{48\sqrt{e}}{\alpha^2} \frac{\rho_*^4}{a_*^6}. \tag{6.17}$$

In this case, the density grows to infinity at two points:

$$\rho_{wb} = \rho_* / (2\sqrt{3}\alpha), \quad \varphi_{wb}^1 = \pi / 2, \quad \varphi_{wb}^2 = 3\pi / 2, \tag{6.18}$$

which are located symmetrically with respect to the x axis. In the opposite case, when $0 < \alpha < 1$, the maximum absolute value of the function in (6.14) $F_{\max}(\rho,\varphi) = (3/16)(a_* / \rho_*)^6 \rho_*^2 / (9\sqrt{e})$ is reached at $\varphi = 0$ and $\varphi = \pi$ at a distance from the axis $\rho_0 = \rho_* / (2\sqrt{3})$. This corresponds to the breaking time

$$\theta_{wb} \approx 48\sqrt{e} \frac{\rho_*^4}{a_*^6}. \tag{6.19}$$

at two points with coordinates

$$\rho_{wb} = \rho_* / (2\sqrt{3}), \quad \varphi_{wb}^1 = 0, \quad \varphi_{wb}^2 = \pi, \tag{6.20}$$

which are located symmetrically about the y-axis.

From the obtained results, it follows that the breaking of axially symmetric electronic oscillations is unstable in shape, that is, even a small deviation leads to a significant change in the structure of the electron density. In contrast to axially symmetric oscillations, where the breaking took place on a circle of radius $\rho_{wb} = \rho_* / (2\sqrt{3})$ in the case considered $\alpha \neq 1$, the breaking of plasma oscillations occurs only at two points (see (6.18) and (6.20)) located at some distance from the axis $\rho = 0$.

6.3. Difference scheme

In numerical simulation, the desired functions are found in a bounded domain

$$\Omega = \{|x| \leqslant d, \quad |y| \leqslant d, \quad 0 \leqslant \theta \leqslant \theta_{\max}\},$$

and the boundary conditions for the functions E_x, E_y, P_x, P_y, B_z, unless otherwise stated, are the conditions for localizing oscillations with respect to spatial variables, i.e. equal to zero at an arbitrary time θ, if one of the x or y coordinates coincides in modulus with d.

As a basis for time sampling, we use the usual leapfrog scheme. Let τ be a time step, then we will assign the values $\mathbf{E} = (EX, EY, 0)$, N to 'integer' moments of time $\theta_j = j\tau$ $(j \geq 0$ – integer), and $\theta_{j\pm1/2}-\mathbf{B} = (0, 0, B)$ to 'half whole', $\mathbf{P} = (PX, PY, 0)$. The choice of the corresponding time moment for the function value will be denoted by the superscript. From this point on, the notation for the desired functions is slightly changed to make room for subscripts.

For discretization in space, we will use the so-called shifted grids, that is, for numerically solving two-dimensional equations, each of the functions N, EX, EY, PX, PY and B is determined on a special finite set of nodes. In the case under consideration, we need displaced grids D_i $(0 \leq i \leq 3)$ in the plane of variables (x, y). Introducing the notation $h_x = d/M_x$, $h_y = d/M_y$, where M_x, M_y are the numbers of nodes in the first quadrant with respect to the variables x, y, respectively, we define:

$$D_0 = \{(x_k, y_l): x_k = kh_x, |k| \leqslant M_x; y_l = lh_y, |l| \leqslant M_y\},$$
$$D_1 = \{(x_k, y_l): x_k = (k+1/2)h_x, -M_x \leqslant k \leqslant M_x - 1;$$
$$y_l = lh_y, |l| \leqslant M_y\},$$
$$D_2 = \{(x_k, y_l): x_k = kh_x, |k| \leqslant M_x;$$
$$y_l = (l+1/2)h_y, -M_y \leqslant l \leqslant M_y - 1\},$$
$$D_3 = \{(x_k, y_l): x_k = (k+1/2)h_x, -M_x \leqslant k \leqslant M_x - 1;$$
$$y_l = (l+1/2)h_y, -M_y \leqslant l \leqslant M_y - 1\}.$$

The result of discretization in space will be denoted by subscripts, and the function N is determined on the grid D_0, EX, PX on D_1, EY, PY – on D_2, B – on D_3, respectively.

If, in modelling the plasma processes we use spatially separated grids and stepping over time, they traditionally refer to work [170], although it should be noted that discrete analogs of the main differential operators were introduced in [70] on shifted grids.

We give a description of the computational algorithm as a whole, assuming that at the beginning of the time step on the grids defined above, we know the values:

$$PX^{j-1/2}, \quad PY^{j-1/2}, \quad B_{k,l}^{j-1/2}, \quad EX_{k,l}^j, \quad EY_{k,l}^j, \quad N_{k,l}^j.$$

6.3.1. Difference equations in the internal nodes of the grid

We describe the sequence of calculations.
Step 1. First recalculate the magnetic field

$$\frac{B_{k,l}^{j+1/2} - B_{k,l}^{j-1/2}}{\tau} = -\left(\frac{EY_{k+1,l}^{j} - EY_{k,l}^{j}}{h_x} - \frac{EX_{k,l+1}^{j} - EX_{k,l}^{j}}{h_y} \right).$$

Step 2. Calculation of the projections of the momentums and the Lorentz factor consists of two stages: a predictor and a corrector. This is done to ensure second-order accuracy in time on smooth solutions with nonlinear terms taken into account (primarily for *PX* and *PY*).

The predictor. Calculate on the layer $(j - 1/2)$ the function-sum of the squares of the momentus on the grid D_0 and the corresponding value of γ:

$$S_{k,l}^{j-1/2} = (PX_{k,l}^2 + PX_{k-1,l}^2 + PY_{k,l}^2 + PY_{k,l-1}^2)) / 2,$$

$$\gamma_{k,l}^{j-1/2} = \sqrt{1 + S_{k,l}^{j-1/2}}.$$

Then we integrate the equations according to the scheme

$$\frac{\overline{PX}_{k,l}^{j+1/2} - PX_{k,l}^{j-1/2}}{\tau} = -\left(EX_{k,l}^{j} + \frac{S_{k+1,l}^{j-1/2} - S_{k,l}^{j-1/2}}{h_x(\gamma_{k+1,l}^{j-1/2} + \gamma_{k,l}^{j-1/2})} \right),$$

$$\frac{\overline{PY}_{k,l}^{j+1/2} - PY_{k,l}^{j-1/2}}{\tau} = -\left(EY_{k,l}^{j} + \frac{S_{k,l+1}^{j-1/2} - S_{k,l}^{j-1/2}}{h_y(\gamma_{k,l+1}^{j-1/2} + \gamma_{k,l}^{j-1/2})} \right).$$

These values $\overline{PX}$ and $\overline{PY}$ have only the first order of accuracy in τ due to the use of nonlinear terms from the bottom layer over time. Therefore, to obtain the second order of accuracy on smooth solutions, a step-corrector is required.

Corrector. Using the calculated $\overline{PX}_{k,l}^{j+1/2}$ and $\overline{PY}_{k,l}^{j+1/2}$, we calculate on the layer j the function-sum of squares of momentums on the grid D_0 and the corresponding value γ:

$$\overline{S}_{k,l}^{j+1/2} = (\overline{PX^2}_{k,l} + \overline{PX^2}_{k-1,l} + \overline{PY^2}_{k,l} + \overline{PY^2}_{k,l-1})) / 2,$$

$$S_{k,l}^{j} = (\overline{S}_{k,l}^{j+1/2} + S_{k,l}^{j-1/2}) / 2, \quad \gamma_{k,l}^{j} = \sqrt{1 + S_{k,l}^{j}}.$$

Then we integrate the same equations a second time (as a result, we have an analogue of the Euler scheme with recalculation [20])

$$\frac{PX_{k,l}^{j+1/2} - PX_{k,l}^{j-1/2}}{\tau} = -\left(EX_{k,l}^{j} + \frac{S_{k+1,l}^{j} - S_{k,l}^{j}}{h_x(\gamma_{k+1,l}^{j} + \gamma_{k,l}^{j})} \right),$$

$$\frac{PY_{k,l}^{j+1/2} - PY_{k,l}^{j-1/2}}{\tau} = -\left(EY_{k,l}^{j} + \frac{S_{k,l+1}^{j} - S_{k,l}^{j}}{h_y(\gamma_{k,l+1}^{j} + \gamma_{k,l}^{j})} \right).$$

At the end of the step, we calculate (refine) the value of γ at the moment $j + 1/2$ on the basis of the quantities $PX_{k,l}^{j+1/2}$ and $PY_{k,l}^{j+1/2}$ just obtained:

$$S_{k,l}^{j+1/2} = (PX_{k,l}^2 + PX_{k-1,l}^2 + PY_{k,l}^2 + PY_{k,l-1}^2)) / 2,$$

$$\gamma_{k,l}^{j+1/2} = \sqrt{1 + S_{k,l}^{j+1/2}}.$$

Step 3. Finally, you can recalculate the electric field based on the formulas

$$\frac{\overline{EX}_{k,l}^{j+1} - EX_{k,l}^{j}}{\tau} = 2\frac{PX_{k,l}^{j+1/2} N_{k,l}^{j}}{\gamma_{k+1,l}^{j+1/2} + \gamma_{k,l}^{j+1/2}} + \frac{B_{k,l}^{j+1/2} - B_{k,l-1}^{j+1/2}}{h_y},$$

$$\frac{\overline{EY}_{k,l}^{j+1} - EY_{k,l}^{j}}{\tau} = 2\frac{PY_{k,l}^{j+1/2} N_{k,l}^{j}}{\gamma_{k,l+1}^{j+1/2} + \gamma_{k,l}^{j+1/2}} - \frac{B_{k,l}^{j+1/2} - B_{k-1,l}^{j+1/2}}{h_x},$$

where

$$N_{k,l}^{j} = 1 - \left(\frac{EX_{k,l}^{j} - EX_{k-1,l}^{j}}{h_x} + \frac{EY_{k,l}^{j} - EY_{k,l-1}^{j}}{h_y} \right). \tag{6.21}$$

At this stage, in order to increase the order of accuracy, as in step 2, the predictor-corrector scheme is applied to the multiplier $N_{k,l}$ in non-stationary equations. Here it is meant that at the stage the predictor is calculated for layer j, as indicated in (6.21), and at the stage corrector – already as a half-sum for layers: main j and intermediate

$j + 1$, i.e. on the basis of the calculated $\overline{EX}_{k,l}^{j+1}$ and $\overline{EY}_{k,l}^{j+1}$ by the formula (6.21) we have $\bar{N}_{k,l}^{j+1}$ and then $N_{k,l}^{j+1/2} = (N_{k,l}^{j} + \bar{N}_{k,l}^{j+1})/2$. Further calculations practically reproduce the step – predictor:

$$\frac{EX_{k,l}^{j+1} - EX_{k,l}^{j}}{\tau} = 2\frac{PX_{k,l}^{j+1/2} N_{k,l}^{j+1/2}}{\gamma_{k+1,l}^{j+1/2} + \gamma_{k,l}^{j+1/2}} + \frac{B_{k,l}^{j+1/2} - B_{k,l-1}^{j+1/2}}{h_y},$$

$$\frac{EY_{k,l}^{j+1} - EY_{k,l}^{j}}{\tau} = 2\frac{PY_{k,l}^{j+1/2} N_{k,l}^{j+1/2}}{\gamma_{k,l+1}^{j+1/2} + \gamma_{k,l}^{j+1/2}} - \frac{B_{k,l}^{j+1/2} - B_{k-1,l}^{j+1/2}}{h_x}.$$

At the end of the time loop, the 'real' (and not intermediate!) electron density $N_{k,l}^{j+1}$ is calculated again using formula (6.21).

Note that in modelling by th the particle-in-cell (PIC) method, schemes of the proposed type are usually used to calculate fields [167]. It is also known [170] that the stability condition in the case under consideration has the asymptotics $\tau = O\left(\sqrt{h_x^2 + h_y^2}\right)$, which fully agrees with the computational experiments given below.

6.3.2. Implementation of the artificial boundary conditions

In Section 3.6, devoted to the construction of the artificial boundary conditions, it was noted that their various combinations are possible, essentially relying on the specific formulation of the problem.

The simplest variant of the boundary conditions is the 'full damping of oscillations':

$$P_x = P_y = E_x = E_y = B_z = 0,$$

when the value of any independent variable (or both at once) coincides with the boundary of the domain $d = 4.5\rho_*$. This case, due to the obvious implementation, does not require comments.

More meaningful is the combined set of boundary conditions for the problem in question. We give a description of it only on one of the four parts of the boundary (for short); fix it:

$$x = d, \quad -d \leqslant y \leqslant d, \quad d = 2.0 * \rho_*.$$

In terms of the differential formulation of the problem, the set of artificial boundary conditions has the following form:

$$\frac{\partial P_x}{\partial \theta} + E_x = 0, \quad P_y = 0, \quad \frac{\partial E_x}{\partial \theta} = \frac{P_x N}{\gamma} + \frac{\partial B_z}{\partial y}, \quad E_y = 0,$$

that is, for P_x, the 'truncated' equation to the linear terms is used, the tangent components of the vector functions are simply zeroed, and for E_x an approach based on 'deterioration' of approximation is used. On the rest of the boundary the conditions are formulated in a similar way.

Recall that the usual leapfrog scheme is used as a basis for time discretization. Let τ be a time step, then the 'integral' time instants $\theta_j = j\tau$ ($j \geq 0$ is an integer) include the values $\mathbf{E} = (EX, EY, 0)$, N, and the 'half-integer' $\theta_{j\pm1/2}$ — $\mathbf{B} = (0, 0, B)$, $\mathbf{P} = (PX, PY, 0)$. The selection of the appropriate time for the function value is indicated by the superscript. We will also slightly change the notation of the functions we are looking for to make room for subscripts.

For discretization in space we use the so-called shifted grids, described above that is, for the numerical solution of two-dimensional equations, each of the functions N, EX, EY, PX, PY and B is determined on a special finite set of nodes. The result of discretization in space is denoted by subscripts.

Since the applied difference scheme is explicit, we assume that at the beginning of the time step on the grids defined above, we know the values:

$$PX^{j-1/2}, \quad PY^{j-1/2}, \quad B_{k,l}^{j-1/2}, \quad EX_{k,l}^{j}, \quad EY_{k,l}^{j}, \quad N_{k,l}^{j}.$$

We describe step by step the numerical implementation of the used combination of the artificial boundary conditions; in the index form, the discussed part of the boundary has the form: $k = M_x$, $-M_y \leq l \leq M_y$.

Step 1. First, we calculate the magnetic field with the following values of the indices:

$$k = M_x - 1, -M_y \leqslant l \leqslant M_y - 1,$$

$$\frac{B_{k,l}^{j+1/2} - B_{k,l}^{j-1/2}}{\tau} = -\left(\frac{EY_{k+1,l}^{j} - EY_{k,l}^{j}}{h_x} - \frac{EX_{k,l+1}^{j} - EX_{k,l}^{j}}{h_y} \right).$$

The formulas for the magnetic field at the boundary do not change as compared with the formulas inside the domain.

Step 2. Integrating the 'truncated' equations for the normal component PX and zeroing the tangent component PY:

$$\frac{PX_{k,l}^{j+1/2} - PX_{k,l}^{j-1/2}}{\tau} = -EX_{k,l}^{j}$$

at $k = M_x - 1, -M_y + 1 \leqslant l \leqslant M_y - 1,$

$$PY_{k,l}^{j+1/2} = 0 \quad \text{at} \quad k = M_x, \quad -M_y + 1 \leqslant l \leqslant M_y - 1,$$

and then, using the values for the projections of the momentums $PX_{k,l}^{j+1/2}$ and $PY_{k,l}^{j+1/2}$, we calculate the formula $\gamma_{k,l}^{j+1/2}$ using formulas that do not change compared to the formulas inside the domain.

Step 3. We recalculate the boundary values for the electric field by the formulas

at $k = M_x - 1, -M_y + 1 \leqslant l \leqslant M_y - 1,$

$$\frac{EX_{k,l}^{j+1} - EX_{k,l}^{j}}{\tau} = \frac{PX_{k,l}^{j+1/2} N_{k,l}^{j}}{\gamma_{k,l}^{j+1/2}} + \frac{B_{k,l}^{j+1/2} - B_{k,l-1}^{j+1/2}}{h_y},$$

at $k = M_x, \quad -M_y + 1 \leqslant l \leqslant M_y - 1, \quad EY_{k,l}^{j+1/2} = 0.$

The formulas used for the calculation in the internal nodes of the domain use more complex expressions: $\left(N_{k,l}^{j} + N_{k+1,l}^{j}\right)/2$ instead of $N_{k,l}^{j}$ and $\left(\gamma_{k,l}^{j+1/2} + \gamma_{k+1,l}^{j+1/2}\right)/2$ instead of $\gamma_{k,l}^{j+1/2}$, respectively. Simplified expressions in the boundary nodes reduce the formal order of approximation, as described in section 3.6.4, but the nodes of the template used do not go beyond the limits of the computational domain.

It should be noted that the recalculation of the boundary values is carried out within each computational step immediately after the determination of the corresponding functions at the interior points of the domain.

6.4. Numerical experiments

6.4.1. General remarks

Note that the considered oscillation breaking problem (6.2), (6.3) is rather complicated from the computational point of view. First, very large values of electron density appear in the hydrodynamic description in the Eulerian coordinates when approaching the moment of breaking in the vicinity of the points of singularity. Secondly, the very value of the coordinate in the time of oscillation failure is very sensitive to the input data: in the weakly non-linear approximation it is inversely proportional to the third power (cube!) of the initial amplitude. This means that determining the physical parameters of a computationally accessible version is already a difficult task. Third, the radial (in the axially symmetric case) breaking coordinate in a moderately nonlinear mode, that is, even when the electron density perturbation exceeds the background value only by an order of magnitude, is about 1–2% of the characteristic size of the computational domain. In other words, in order to adequately display the process, thousands of points along each spatial coordinate are required, even under the condition that the desired functions are smooth and bounded.

Considering the above, for calculating plane two-dimensional relativistic electron oscillations in a cold plasma a special difference scheme was constructed on spatially separated grids described in the previous section. Special publications [76, 122] are devoted to its testing in sequential and parallel implementations. It should be noted that the scheme used on smooth solutions has the second order of approximation both in spatial variables and in time. In this case, the spatial approximation is achieved at the expense of shifted grids, and the temporal approximation is achieved at the expense of computational methods such as predictor–corrector. It is important that the scheme admits realization by explicit formulas, and the condition for its stability, following their principle of frozen coefficients [20, 41], has an asymptotic Courant condition, that is, $\tau = O\left(\sqrt{h_x^2 + h_y^2}\right)$. Here we use the usual notation for steps in spatial variables and time.

The need to use fine grids to simulate breaking

$$h_x M_x = h_y M_y = d, \quad M_x, M_y \approx 10^3 \div 10^4$$

in combination with the explicit design of the design scheme, it is necessary to use the enormous computing resources currently available only within the framework of modern high-performance computing clusters. In the overwhelming majority of such systems, a hybrid parallel architecture is implemented, which implies combining a set of multicore nodes with distributed memory through a high-performance channel. These circumstances determined the need to write a hybrid parallel version of the program. Moreover, the fact that the scheme used is explicit, as well as the regular structure of the computational grid, made it possible to implement an efficient parallel code [76] without significant changes to the sequential version of the program [122].

In the program used for the calculations, hybridity consisted in the presence of actually two levels of parallelization: between cluster nodes (implemented using the MPI library), and also within each calculation node (implemented using pthreads) [24]. In addition, one MPI process was launched on each cluster node. Due to the use of a uniform rectangular grid, the decomposition problem was solved simply by cutting the grid domain into identical bands: the node with number i worked with the subgrid

$$(k,l) \in \left[\frac{2M_x i}{N_{\text{tot}}} - (M_x + 1); \frac{2M_x(i+1)}{N_{\text{tot}}} - (M_x - 1)\right] \times \left[-M_y; M_y\right],$$

where N_{tot} is the total number of nodes. In turn, within each node the distribution of the problem between the cores was also uniform.

In addition, to optimize the memory access time with non-uniform memory architecture (NUMA-systems [54]), an algorithm was implemented that minimizes the number of stream accesses to non-local memory of the NUMA-node on which it operates. This was realized due to the fact that, firstly, for each computational flow it was linked to the NUMA node, and secondly, the initialization of the memory with which this thread mainly operates occurred on the NUMA node on which it works. It should be noted that such a measure made it possible to actually double the scalability when running on a large number of cluster nodes. Supercomputer numerical experiments were carried out on the 'Chebyshev' supercomputer at the SKIF Moscow State University [1], consisting of nodes with two quad-core Intel Xeon E5472 processors connected by an InfiniBand DDR network.

6.4.2. Calculations with circular symmetry

We give a description of the computational experiment that generalizes the one-dimensional calculations from Chapter 4.

We fix the parameters of the initial data $a_* = 0.365$, $\rho_* = 0.6$, the computational domain $d = 4.5\rho_* = 2.7$ and the grid steps $h_x = h_y = 1/1600$, $\tau = 1/8000$, and then we calculate by time until

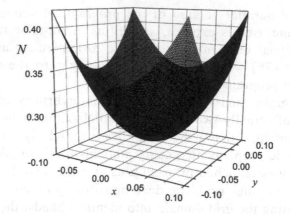

Fig. 6.1. The electron density distribution at the initial time.

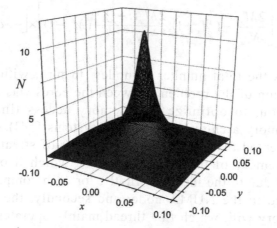

Fig. 6.2. The first electron density maximum located in the centre of the domain.

$\theta \approx 40$, corresponding to the formation of the second off-axis maximum density.

The oscillations begin with a certain initial perturbation of the electron density function shown in Fig. 6.1. It is easy to see that the function depicted has a circular (axial) symmetry, which corresponds to the equally symmetrical intensity of a laser pulse, which has an additional Gaussian distribution in the transverse direction. An excess of positive charge at the origin of the coordinates leads to the movement of electrons in the direction of the centre of the domain, which, after half a period of oscillation, generates another distribution of the density function, shown in Fig. 6.2. Note that the concentration of the electrons in the centre of the domain at times of regular (periodic) maxima can many times exceed the equilibrium value. Here are calculations of the fluctuations of low intensity, when the electron density is only about 10 times the maximum of its background value.

If the plasma oscillations would keep their spatial form in time, then the presented electron density images would regularly change each other through each half of the period, generating a strictly periodic sequence of extremes with constant amplitudes in the centre of the domain. However, in the process of oscillation propagation, the phase front is gradually warped; therefore, after a certain number of periods, a fundamentally new structure of electron density is formed – with a circular maximum comparable in size to regular maxima in the centre. In the calculated version, Fig. 6.3, this happened in the sixth period of free oscillations. Note that for clarity the figure shows not the whole region, but only its quarter. We emphasize the important differences of the new structure: the local in time and the global in space, the density maximum appears on the period at another time (shortly after the maximum at the centre), in another place (not at one point, but at the circle) and at significant scales (comparable to regular minimum).

Further in the process of oscillations, this structure is reproduced with a monotonously increasing amplitude. However, after its appearance, it is first converted into a semblance of a regular central density minimum, then its regular central maximum is formed, and after it, another circular maximum. The electron density function corresponding to this point in time is shown in Fig. 6.4 also in a quarter of the domain. The value of this maximum density already exceeds the regular central value by about two times.

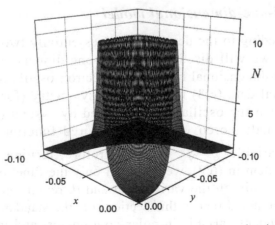

Fig. 6.3. The first circular maximum electron density.

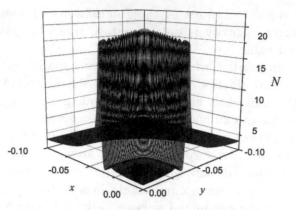

Fig. 6.4. The second circular maximum of the electron density, which occurred approximately after the period after the first circular maximum.

Recall that this off-axis maximum is a precursor of breaking with the considered parameters. In the next period of oscillations, in place of a circular extremum, infinite values of electron density are formed. This means that the solution of the problem gave rise to a 'step' of the electric field, whose divergence has a singularity.

The picture described is in excellent agreement with one-dimensional calculations, carefully carried out within the framework of various models [45, 46, 100]. In addition, it should be noted that system (6.2) has, as partial solutions, zero-dimensional axial solutions, for which not only necessary and sufficient conditions of existence were obtained but also analytical formulas in section 2.2 (see also [102]).

6.4.3. Quasi-one-dimensional model

Before proceeding to the description of essentially two-dimensional calculations, we will discuss the quasi-one-dimensional model of breaking two-dimensional relativistic electron oscillations.

If the initial data (6.3) have circular symmetry (for example, as for $\alpha = 1$), then the oscillations described by system (6.2) become one-dimensional; moreover, all the required functions depend on the spatial variable $\rho = \sqrt{x^2 + y^2}$. Therefore, turning to the polar coordinate system in (6.2), we find that only the functions E_ρ, P_ρ, N, which depend only on the variables ρ and θ, remain nonzero.

It should be noted that in this section ρ is the standard designation of the independent variable in polar coordinates, and in section 6.2

– simply convenient in the sense of the notation for an auxiliary parameter characterizing the normalized distance from the origin.

Omitting the subscript ρ, the equations can be written as

$$\frac{\partial P}{\partial \theta} + E + \frac{1}{2\gamma}\frac{\partial P^2}{\partial \rho} = 0, \quad \gamma = \sqrt{1+P^2},$$

$$\frac{\partial E}{\partial \theta} = N\frac{P}{\gamma}, \quad N - 1 - \frac{1}{\rho}\frac{\partial \rho E}{\partial \rho}. \tag{6.22}$$

The required functions are sought in the region bounded in space $|\rho| \leq d$. As above, the boundary conditions for E and P are localization conditions for oscillations in space, that is, they are equal to zero at an arbitrary time θ, if the coordinate ρ in modulus coincides with d. The initial conditions, similar to (6.3), will be supplied with the parameter β:

$$E(\rho, \theta = 0) = \left(\frac{a_*}{\rho_*}\right)^2 \beta\rho\exp^2\left\{-\frac{\beta\rho^2}{\rho_*^2}\right\}, \quad P(\rho, \theta = 0) = 0. \tag{6.23}$$

The particle trajectory is determined by the expression $\rho = \rho_0 + R\,(\rho_0,\ \theta)$, where ρ_0 is the initial equilibrium position of the particle, which does not lead to an electric field, $R(\rho_0,\ \theta)$ is the displacement relative to the equilibrium position, which in the cylindrical one-dimensional case is related to electric field by the formula $E = \frac{1}{2}\frac{\rho^2 - \rho_0^2}{\rho}$. In the case under consideration, we have an asymptotic formula (see Section 4.1, as well as [46, 123])

$$\rho = \rho_0 + R_0\cos\left(1 + \frac{R_0^2}{12\rho_0^2} - \frac{3R_0^2}{16}\right)\theta, \quad R_0 = R(\rho_0, \theta = 0). \tag{6.24}$$

Here, the oscillation frequency $\omega = 1 + \dfrac{R_0^2}{12\rho_0^2} - \dfrac{3R_0^2}{16}$ also has an amendment to the fundamental frequency equal to unity. In this case, the correction quadratically depends on the oscillation amplitude R_0, and the difference in the frequencies of the neighbouring particles further leads to breaking of the oscillations. Note that in the case of cylindrical relativistic oscillations, the correction consists already of two terms: the first is a consequence of axial symmetry, that is, the geometric properties of the problem, and the second is determined by the relativistic factor (see [46, 123]).

We define the quasi-one-dimensional construction of oscillations as follows. Let us introduce in the parameter β a dependence on the polar angle φ, i.e. we will assume that a sufficiently smooth 2π-periodic function $\beta = \beta(\varphi)$ is defined such that $1 \leq \beta(\varphi) \leq \alpha$, and

$$\beta(0) = 1, \quad \beta(\pi / 2) = \alpha, \quad \beta(\pi) = 1, \quad \beta(3\pi / 2) = \alpha$$

and for the above values of the angle, its derivative vanishes. The simplest example of such a function is a periodic polynomial or trigonometric spline $\beta(\varphi)$, varying from unity to α on each segment $[0, \pi/2]$, $[\pi/2, \pi]$, $[\pi, 3\pi/2]$, $[3\pi/2, 2\pi]$ and continuing evenly when moving from one segment to another.

Further, for each $0 \leq \varphi \leq 2\pi$, in accordance with the statement (6.22), (6.23), we define the functions $E_{\beta(\varphi)}$, $P_{\beta(\varphi)}$, $N_{\beta(\varphi)}$. These functions will satisfy the original two-dimensional equations (6.2) and initial conditions (6.3) up to terms of the form $\dfrac{1}{\rho} \dfrac{\partial}{\partial \varphi}$, therefore, for small deviations from circular symmetry, that is, with $|\alpha-1| \ll 1$, the most significant deviations of the quasi-one-dimensional model should be expected in the vicinity of the origin. In turn, the effect of breaking oscillations always occurs at a certain distance from zero, which is proportional to the parameter ρ_*. This means that for relativistic oscillations, i.e. in the case of $\rho_* \gg 1$, the deviation of the quasi-one-dimensional model from the exact solution (6.22), (6.23) in the vicinity of the breaking zone is small in quantitative terms, therefore, the breaking process must be reproduced itself efficiently with sufficient completeness.

Recall for illustrative purposes a typical form of the desired functions, which characterize the effect of breaking one-dimensional cylindrical relativistic oscillations. Fix the parameters in (6.23) as follows:

$$a_* = 1.75, \quad \rho_* = 3.0, \quad \beta = 1.05;$$

We define the boundary of the computational domain as $d = 4.5\rho_* = 13.5$, choose the grid with constant steps: $h = 1/200$, $\tau = 1/500$. The most reliable way to obtain an approximate solution of the problem (6.22), (6.23) is the calculation by the particle method (in Lagrangian variables), described in sec. 4.4 (see also [46]). In this case, finding an approximate solution is reduced to independent integration of a system of ordinary differential equations that have smooth solutions.

Consider in Fig. 6.5 the graph of electron density, indicated by a solid line. It clearly shows the following trend: over time, there is a gradual formation of the absolute maximum density, located off-axis and comparable in magnitude with the axial ones. At first, the oscillations are of a regular nature, with each half of the period a global maximum of the density in the domain, approximately equal to 8.4, located strictly on the axis of symmetry. After the fifth regular (axial) maximum, approximately at $\theta \approx 28.2$, a new structure arises – an off-axis maximum of the electron density. At the same time, oscillations of axial maxima continue to be observed with the same periodicity. The off-axis maximum, in turn, at an instant of time $\theta \approx 34.7$, increases in magnitude by about three times and in the next period, at $\theta_{wb} \approx 40.3$, the singularity of the electron density appears in its place. The radial coordinate of the breaking point is $\rho_{wb} \approx 0.19$. The behaviour of the off-axis maximum of the electron density is clearly shown in Figs. 6.6 and 6.7. Figure 6.6 depicts the spatial distribution of density at the moment $\theta \approx 34.7$, when it was already fully formed and became in absolute value a substantially more regular axial maximum. The density graph in Fig. 6.6. is a consequence of the distributions of the velocity V and the electric field E shown in Fig. 6.7. Note that, in the vicinity of the density maximum, the velocity function tends to jump in the derivative, and the electric field function takes on a stepwise character. It is these qualitative characteristics of V and E that ensure oscillation breaking at the time $\theta_{wb} \approx 40.3$. It is important to note that the breaking has

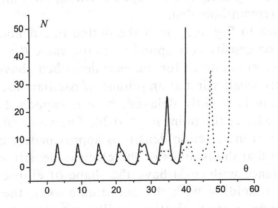

Fig. 6.5. The dynamics of the maximum electron density in the domain: the solid line for $\beta = 1.05$, the dotted line for $\beta = 1$.

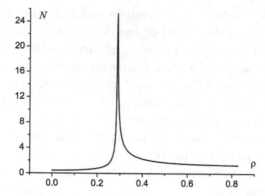

Fig. 6.6. Spatial distribution of electron density at the moment of formation of the second off-axis maximum.

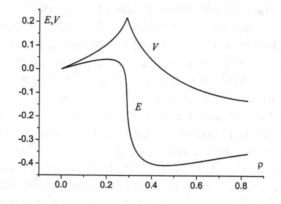

Fig. 6.7. Spatial distribution of velocity and electric field at the moment of formation of the second off-axis maximum.

the character of a 'gradient catastrophe', that is, the functions V and E themselves remain bounded.

Let's go back to Fig. 6.5, on it the dotted line denotes a similar graph of electron density corresponding to the value $\beta = 1$. It has the same qualitative structure as for the case described above. However, due to a slightly smaller initial amplitude of oscillations, the process of their breaking is slightly delayed: here it happened at the time moment $\theta_{wb} \approx 53.1$ at the point $\rho_{wb} \approx 0.24$. The value of the regular axial maximum of the electron density is approximately equal to 6.2.

Considering that the isolines of the electron density at the initial time in accordance with (6.3) have the shape of ellipses with axes of symmetry coinciding with the coordinate axes, the quasi-one-dimensional model predicts electron oscillations along the coordinate axes depending only on the distance from the origin of coordinates.

In other words, the electron oscillations along the OY axis should be well described by solving the problem (6.22), (6.23) with $\beta = 1.05$, and the electron oscillations along the OX axis should be solved by the problem (6.22), (6.23) with $\beta = 1$. The main difference a two-dimensional model should be observed in the vicinity of the origin, since even with a small deviation from circular symmetry, the terms of the form $\dfrac{1}{\rho}\dfrac{\partial}{\varphi}$ in complete equations significantly influence the solution. For example, determine the continuity of the electron density function at the origin.

In accordance with the graphs in Fig. 6.5 the quasi-one-dimensional model predicts that the breaking of oscillations must occur necessarily on the OY axis (i.e., at the polar angle $\pi/2$ and $3\pi/2$). In this case, the breaking on the OX axis simply does not have time to be realized, due to the lack (smallness) of the corresponding shift in the oscillation frequency. In addition, it can be predicted that the moment of breaking two-dimensional oscillations will occur in the interval [40.3, 53.1], i.e. will be located strictly between the times of one-dimensional breaking effects. A similar forecast is quite acceptable for other important characteristics of the process of two-dimensional oscillations: the maximum of the electron density at the origin should take values from the interval [6.2, 8.4], and the radial coordinate of the breaking – from the interval [0.19, 0.24].

We note once again that such forecasts, which do not take into account the angular changes in the solution of the two-dimensional problem, are permissible only outside a certain neighbourhood of the origin and for large sizes of the oscillation region.

6.4.4. Small deviation from circular symmetry

To illustrate the conclusions of the asymptotic analysis of the oscillations and the prediction for the quasi-one-dimensional model, we present the results of calculations of problem (6.2), (6.3). At the same time, the parameters of the initial data are selected in such a way that the breaking is determined by the relativistic effect, and the deviation from axial symmetry is insignificant:

$$a_* = 1.75, \quad \rho_* = 3.0, \quad \beta = 1.05;$$

the boundary of the computational domain is given by the value

$d = 4.5\rho_* = 13.5$, the grid is selected with constant steps $M_x = M_y = 4321$, $\tau = 1/32\,000$.

Consider in Fig. 6.8 plot of electron density, indicated by a solid line. It clearly shows the following trend: over time, there is a gradual formation of the absolute maximum density, located outside the origin of coordinates on the OY axis and comparable in magnitude with the axial ones. At first, the oscillations are of a regular nature, with each half of the period a global maximum of density in the domain, approximately equal to 7.0, located strictly at the origin of the coordinates. After the fifth regular maximum, at about $\theta \approx 28.0$, a new structure arises – the electron density maximum, located outside the origin. It should be noted that a similar effect has not yet emerged on the OX axis (dashed line): the time has not yet come. In this case, oscillations of the electron density maxima continue to be observed with the same periodicity.

After about a period, the arisen irregular maximum begins to increase sharply in absolute value. In addition, on the axis OX (dashed line) appears the maximum electron density located outside the origin. This is clearly seen in Fig. 6.8 at the time $\theta \approx 34.5$.

After about a period, the density maximum on the OY axis increases in magnitude by about three times, and a period later — at $\theta_{wb} \approx 42$ – the singularity of the electron density appears in its place. It is easy to see that the oscillations along the OX axis (dashed line) almost completely repeat the oscillations along the OY axis, but with a smaller amplitude. Such a situation corresponds exactly to the prediction for a quasi-one-dimensional model: for $\alpha = 1$, both graphs merge into one because of the circular symmetry of the oscillations.

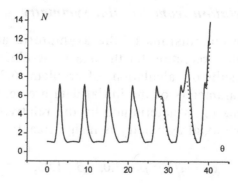

Fig. 6.8. The dynamics of the maximum electron density along the axes of symmetry: the solid line for the axis OY, the dotted line for the axis OX.

For clarity, the qualitative behaviour of two-dimensional oscillations will give two illustrations: the spatial density distribution at different points in time.

Figure 6.9 shows the electron density surface corresponding to the time $\theta \approx 15.1$. For convenience, there is only a 'quarter' domain – a non-negative quadrant, by virtue of the evenness of the considered function with respect to both coordinate axes. Here, the maximum of the electron domain is regular, that is, arising periodically and located strictly at the origin of coordinates.

Figure 6.10 shows the surface of the electron density corresponding to the time $\theta \approx 34.5$. Here, the density has pronounced maxima both along the *OY* axis and along the *OX* axis. Their absolute values differ markedly, so the nonlinear growth of each of them leads

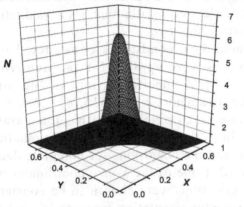

Fig. 6.9. Spatial distribution of the electron density function corresponding to a regular maximum at the origin.

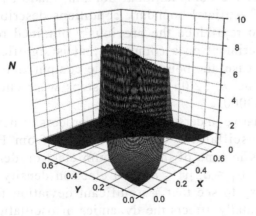

Fig. 6.10. Spatial distribution of the electron density function corresponding to two different maxima outside the origin.

to a singularity at different time moments. In this case, in accordance with the asymptotic theory and the quasi-one-dimensional model, the breaking occurs on the OY axis, i.e. for two values of the polar angle $\varphi = \pi/2$ and $\varphi = 3\pi/2$ at the same distance from the origin.

It should also be noted that the forecast for the quasi-one-dimensional model showed good agreement with the results of two-dimensional numerical simulation for the following parameters: breaking time, coordinate (i.e., distance from the origin) breaking, amplitude of regular oscillations at the origin, etc.

6.4.5. Significant difference from circular symmetry

Let us give a description of the computational experiment, when the initial data differ significantly from axially symmetric ones. In this case, the dynamics of electron oscillations has a qualitative deviation from the cases considered above.

The artificial boundary conditions constructed on the basis of asymptotic analysis were applied for the numerical solution of problem (6.2), (6.3). A specific variant of the initial data is characterized by the parameters $a_* = 0.315$, $\rho_* = 0.6$, $\alpha = 1.5$.

For the first time, a calculation with such parameters was described in [124]. In it, the boundary of the computational domain was set to $d = 4.5\rho_*$ and the conditions for complete damping of the oscillations were used if one of the x or y coordinates was equal to d in modulus. The grid steps were chosen to be constant: $h_x = h_y = 1/3200$, $\tau = 1/32\,000$, the calculation was carried out at the SKIF Moscow State University 'Chebyshev' supercomputer.

The reduction of the computational domain, made by selecting the combination of artificial boundary conditions described above, made it possible to reproduce the previously obtained result of a computational experiment on a computer that is significantly less productive. In this case, the boundary of the computational domain was set by the value $d = 2.0\rho_*$, the grid steps were chosen: $h_x = h_y = 1/1500$, $\tau = 1/3000$.

The initial qualitative understanding of the process of development – destruction of oscillations can be obtained from Fig. 6.11, illustrating changes in time of the maximum electron density over the domain $\{(x, y): |x| \leq d, |y| \leq d\}$ and electron density values at the origin. It is easy to see that a significant deviation from axial symmetry fundamentally affects the dynamics of oscillations. First, the electron density maxima, located at the origin of coordinates, first

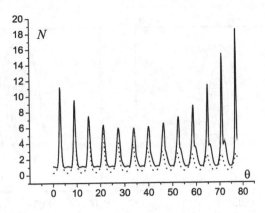

Fig. 6.11. Dynamics of electron density: maximum in the domain (solid line) and at the origin (dashed line),

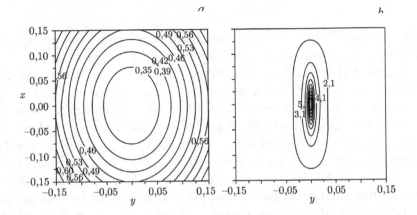

Fig. 6.12. Beginning of oscillations: isolines of the electron density with an interval of half a period. The time moments correspond to the extremes of density in the centre of the domain.

monotonically decrease each period and then stabilize at a very small amplitude, which is about 5 times less than the initial maximum. Secondly, the electron density global maxima coincide in time with local maxima from the origin; at first they decrease, then begin to monotonously increase. Note that even from the uninformative Fig. 6.11, it can be concluded that the breaking of oscillations occurs at a certain distance from the centre of the region, i.e. it has an off-axis character.

Consider the first period of oscillation. Figure 6.12 shows the isolines of the electron density at the initial moment of time and after

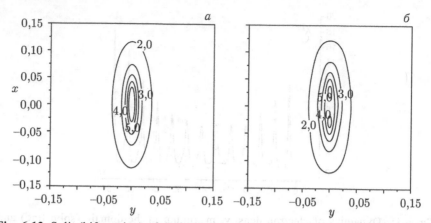

Fig. 6.13. Split (bifurcation) of the central maximum of the electron density: isolines with an interval in the period. Time moments correspond to extreme density values.

half a period. Their elliptical shape leads to a significant quantitative difference from regular oscillations with circular symmetry.

Recall that in the process of development of axially symmetric oscillations, the maximum of the electron density, located in the centre of the domain, appears regularly after a period up to breaking, without changing its shape. When axial symmetry is broken, the central maximum changes from period to period: its absolute value decreases, but at the same time, due to the law of conservation of charge, it begins to 'dissipate', i.e., the electron density perturbation domain begins to increase in size. After several periods of oscillations, this picture becomes distinct, however, it should be noted that while the isolines of the electron density function remain close to ellipses. Due to the dependence of the oscillation frequency on the amplitude, the 'expansion' of the central maximum occurs unevenly, therefore at some time one maximum, local in time and global in space, transforms into two separated in space maxima.

Figure 6.13 shows the isolines of the electron density corresponding to time points $\theta \approx 15.1$ and $\theta \approx 21.4$. On one of them, the maximum electron density 'dissipated' out in space from the origin, and on the next its bifurcation, with the amplitudes of 'twin-maxima' being somewhat less than the amplitude of their one predecessor.

Further, from period to period, a pair of 'twin-maxima', somewhat receding from each other, increases in absolute value: Fig. 6.14 shows the time $\theta \approx 77.1$, when each of a pair of maxima is approximately four times as large as the value of the very first central maximum.

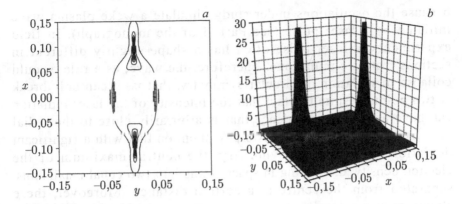

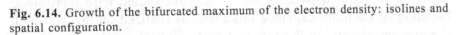

Fig. 6.14. Growth of the bifurcated maximum of the electron density: isolines and spatial configuration.

For clarity, here not only the isolines are shown, but also the spatial configuration of the electron density function. After several periods, the spatial coordinates of a pair of maxima stabilize, and the growth of their absolute values continues.

It should be noted that the process of bifurcation of the central maximum of the electron density and its further development is a consequence of the significant asymmetry (ellipticity) of the initial data, but in no way associated with the usual effect of breaking relativistic oscillations. In accordance with the asymptotic analysis for $\alpha > 1$, the oscillation breaking at points in space is characterized by angles in polar coordinates: $\varphi_{wb}^1 = \pi / 2$, $\varphi_{wb}^2 = 3\pi / 2$. In Fig. 6.14, where the isolines of electron density are indicated, potential singularity points have already appeared at $x = 0$, $|y| \approx 0.04$. Recall that the breaking effect develops rapidly: after one or two periods after the appearance of the first irregular density maxima in their place, a discontinuity of the electric field function and the corresponding infinite value of the electron density are formed. The calculation stops precisely because of the unbounded values of the electron density in the vicinity of the points indicated in section 6.2 (see also [124]).

Thus, based on the asymptotic analysis [124] and the numerical simulation of two-dimensional plane oscillations, we can conclude that the process of development–destruction of axially symmetric oscillations is unstable in shape. Even small deviations from the circular symmetry of the initial perturbations qualitatively change the space–time structure of the solution. This conclusion is important because the oscillations under study simulate a wake plasma wave initiated by a laser pulse (see Part II of the monograph). In field experiments, a laser pulse often has a shape slightly different in sections from a circular one. Therefore, the wave, as a rule, should collapse in violation of circular symmetry, that is, it can only break at two points despite the fact that the intensity of the laser radiation can have a spatial dependence that is arbitrarily close to the radial one. Especially it is necessary to pay attention that with a significant deviation from circular symmetry, the central maximum of the electron density, dissipating in space, turns into two equal maximums, separated from the centre at a certain distance. Moreover, these slowly growing density maxima are not associated with relativistic breaking of oscillations, since this effect is realized on the axis with the greatest change in the gradient of the initial amplitude.

6.5. Bibliography and comments

In the chapter a numerical–asymptotic model based on Eulerian variables is constructed for two-dimensional modeling of the breaking effect of plane relativistic electron oscillations. The main focus of research is the study of the deviation of the solution from circular symmetry.

The asymptotic theory uses the construction of uniformly suitable in time solutions of weakly non-linear equations. For numerical simulation, a special algorithm is used, built on the basis of the finite-difference method and using shifted grids. Difference methods of this type are often used when calculating fields in 'particle-in-cell' methods (PiC). The calculations used serial and parallel versions of the algorithm. In addition, on the basis of numerical and asymptotic studies of axially symmetric one-dimensional relativistic oscillations, a quasi-one-dimensional model of breaking two-dimensional oscillations is proposed, generating two-sided estimates for the main parameters of breaking. Part of the calculations was carried out on the SKIF MSU 'Chebyshev' supercomputer.

The main result of the chapter is the study of the stability of the spatial form of two-dimensional non-linear oscillations at the moment of breaking. It is confirmed that if the initial perturbation is axially symmetric, then the breaking (i.e., the singularity of the electron density function) arises simultaneously on the circle of a certain nonzero radius. A new fact has been established theoretically and numerically: if the sections of the initial perturbation of the electron density are ellipses, then the singularity also arises simultaneously, but only at two points. Moreover, a qualitative change in the final picture of breaking does not depend on the quantitative deviation of the ellipse from the circle: an arbitrary ratio of the ellipse semiaxes, which is different from one, entails the electron density going to infinity only at two points. It should also be noted that if the cause of the breaking of oscillations is primarily the influence of the relativistic factor, then the quasi-one-dimensional model gives a good not only qualitative but also quantitative coincidence with the two-dimensional models.

It seems that the results obtained may be important for the correct interpretation of physical field experiments. It should also be noted that in the vicinity of breaking oscillations one should not neglect the effect of strong electromagnetic fields. It is quite possible that when they appear, heavy ions will also begin to move significantly, thereby changing the pattern of the dynamics of electron density. In other words, from the conducted studies, the relevance of stabilization in the shape of the process of breaking of electronic oscillations follows, which, first of all, requires taking into account the movement of heavier particles [95].

Asymptotic analysis of oscillations given in Section 6.2, was carried out by A.A. Frolov, a parallel version of the program code in Section 6.3 was designed and implemented by S.V. Milyutin.

The results of the chapter were obtained in [75–77, 96, 122, 124].

One should note the fundamental difficulty of using neotravujesfor modelling spatially multidimensional plasma oscillations. This is due to the task of the initial conditions, in the first place, the initial distribution of particles.

Part II

PLASMA WAKE WAVES

7

Introductory information

The chapter discusses various problem statements for hydrodynamic modelling of wake waves in a plasma excited by a short high-power laser pulse. The systems of equations in physical variables, dimensionless variables and variables convenient for carrying out calculations are given. Attention is focused on the basic formulation in nonlinear and linearized variants as on the basis for the design of computational algorithms.

7.1. Source equations

The second part of the book is devoted to laser–plasma interactions, first of all, to the excitation of plasma wake waves as a result of the propagation of a short high-power laser pulse. We describe schematically this process, following [43].

When propagating in a plasma, a pulse pushes electrons out of the domain where it is currently located. In addition to the forces from the side of the pulse, the electrons are affected by the electric field from the plasma ions, which can be considered immobile during the initial consideration because of their large mass. After the pulse has left this domain, only the charge separation field acts on the electrons, seeking to return the electrons to their original position.

Having sped up in this field, the electrons jump through their initial position and begin to oscillate relative to the ions at the so-called *plasma frequency*.

Since the pulse runs through the plasma and all the time pushes out the electrons that are in its path, all the time it starts the plasma oscillations behind it. In this case, the initial phase of these oscillations is different at different points in the path of the pulse. As a result, a charge separation wave is excited, the phase of which propagates through the plasma with a pulse velocity. For such an object there is an established name – *wake wave*. Basically, wake waves are two-dimensional charge density waves that propagate in the direction of the source and are bounded in the transverse direction.

The electric field of this wave in one half of the period is directed towards the propagation of the pulse, and in the other toward the direction of its movement. If an electron with an initial velocity equal to the impulse velocity is placed in that domain of the plasma wave, where the force acting on it from the electric field is directed in the direction of its motion, then the electron, moving along with the wave, will begin to accelerate. Such an accelerator was called a wake wave accelerator [133]. For relativistic particles whose velocity is close to the speed of light, even a small increase in velocity corresponds to a large increase in their energy.

A wake wave excited by a pulse retains its structure only at a certain distance behind the source, and then the wave breaks transferring its energy to the plasma particles. This, on the one hand, limits the size of the region in which the wake wave can be used for controlled acceleration of particles, and on the other hand, leads to uncontrolled trapping and acceleration of electrons in the wake wave breaking domain. The above means the need to study the mechanisms of both regular development and breaking of wake waves.

As a basis for modelling, we will use the equations of two-fluid magnetohydrodynamics of cold plasma (1.5) – (1.12) from Section 1.2.

According to the kinetic model of the propagation of laser pulses in a plasma [153] (see also [10]), equations (1.6), (1.7) should be modified taking into account the slowly varying complex amplitude of the high-frequency laser field $a(\mathbf{x},t)$, $\mathbf{x} \in \mathbb{R}^q$, $q = 1,2,3$ (the so-called *envelope*) as follows:

$$\frac{\partial \mathbf{p}_e}{\partial t} + \left(\mathbf{v}_e \cdot \nabla\right)\mathbf{p}_e = e\left(\mathbf{E} + \frac{1}{c}[\mathbf{v}_e \times \mathbf{B}]\right) - \frac{m_e c^2}{4\gamma}\nabla|a|^2,$$

$$\gamma = \sqrt{1 + \frac{|\mathbf{P}_e|^2}{m_e^2 c^2} + \frac{|a|^2}{2}}.$$

It is easy to see that if the initial functions $\mathbf{B}^0(\mathbf{x}, t)$, $\mathbf{p}_e{}^0(\mathbf{x}, t)$ are related by the relation (1.14)

$$\mathbf{B}^0(\mathbf{x},t) + \frac{c}{e}\operatorname{curl}\mathbf{p}_e{}^0(\mathbf{x},t) = 0,$$

then, in this case it is also possible to write the equation for the electron momentum in a simpler form (1.15). Here, a direct test can establish the validity of the equality for the modified function γ:

$$\left(\mathbf{v}_e \cdot \nabla\right)\mathbf{p}_e = m_e c^2 \nabla\gamma - [\mathbf{v}_e \times \operatorname{curl}\mathbf{p}_e] - \frac{m_e c^2}{4\gamma}\nabla|a|^2.$$

This allows the equation for the momentum of electrons to first lead to the form

$$\frac{\partial \mathbf{p}_e}{\partial t} = e\mathbf{E} - m_e c^2 \nabla\gamma + \frac{e}{c}[\mathbf{v}_e \times \mathbf{A}],$$

where notation for the auxiliary vector field is used

$$\mathbf{A} = \mathbf{B} + \frac{c}{e}\operatorname{curl}\mathbf{p}_e,$$

and then show, as in Statement 1.2.1, that, under condition (1.14), the field $\mathbf{A}$ is identically zero at an arbitrary time.

Consideration of this result is quite important. This is due to the fact that in order to excite a wake wave, the laser pulse hits a stationary plasma, for which relation (1.14) is certainly satisfied. As a result, in the model under consideration, it is possible not only to simplify the form of the equation for the electron momentum, but also to use a simpler equation for the magnetic field.

Consider the open domain $\Omega \subset \mathbf{R}^3$ filled with plasma, and introduce the usual rectangular *OXYZ* coordinate system in it. The *OZ* direction will be considered selected as the direction of propagation of the laser pulse. The pulse profile will be called the function characterizing its intensity in the cross section, i.e., in the *OXY* plane. Let us define a plasma as a mixture of cold, ideal, relativistic electron and non-relativistic ion liquids, then, taking into account the above, the system of hydrodynamic equations describing it together with Maxwell's equations in the quasistatic approximation will look like:

$$\frac{\partial n_e}{\partial t} + \mathrm{div}(n_e \mathbf{v}_e) = 0, \tag{7.1}$$

$$\frac{\partial \mathbf{p}_e}{\partial t} = e\mathbf{E} - m_e c^2 \nabla \gamma, \tag{7.2}$$

$$\gamma = \sqrt{1 + \frac{|\mathbf{p}_e|^2}{m_e^2 c^2} + \frac{|a|^2}{2}}, \tag{7.3}$$

$$\mathbf{v}_e = \frac{\mathbf{p}_e}{m_e \gamma}, \tag{7.4}$$

$$\frac{\partial n_i}{\partial t} + \mathrm{div}(n_i \mathbf{v}_i) = 0, \tag{7.5}$$

$$\frac{\partial \mathbf{v}_i}{\partial t} + (\mathbf{v}_i \cdot \nabla) \mathbf{v}_i = \frac{e_i}{m_i} \left(\mathbf{E} + \frac{1}{c} [\mathbf{v}_i \times \mathbf{B}] \right), \tag{7.6}$$

$$\frac{1}{c} \frac{\partial \mathbf{E}}{\partial t} = -\frac{4\pi}{c} (e n_e \mathbf{v}_e + e_i n_i \mathbf{v}_i) + \mathrm{curl}\, \mathbf{B}, \tag{7.7}$$

$$\mathbf{B} = -\frac{c}{e} \mathrm{curl}\, \mathbf{p}_e, \tag{7.8}$$

where $e(e < 0)$, e_i, m_e, m_i are the charges and masses of electrons and ions, respectively; c is the speed of light; n_e, $\mathbf{p}_e$, $\mathbf{v}_e$ are the concentration, momentum and velocity of electrons; n_i, $\mathbf{v}_i$ is the concentration and velocity of ions; γ is the Lorentz factor; $\mathbf{E}$, $\mathbf{B}$ are the vectors of electric and magnetic fields, a is the slowly varying complex amplitude of the high-frequency laser field (the so-called envelope).

To study the joint dynamics of a plasma and a laser pulse, the system of equations (7.1)–(7.8) should be supplemented by the equation for the envelope a, as described in [133]. However, here the main interest is the excitation of the wake wave by a pulse of a given shape and therefore the corresponding equation is omitted.

Such a situation when simulating laser-plasma interactions is typical and is called 'approximation of a given (sometimes referred to as a stationary) pulse'. This simplification means that the equation for the envelope is not solved in principle, and the influence of laser radiation is transmitted to system (7.1)–(7.8) in the form of a given (fixed) function $a(t, x, y, z)$, which is in the relation (7.3). This is the case that will mainly be considered further.

7.2. The case of an arbitrary pulse velocity

In the section, equations describing the initialization of a three-dimensional axially symmetric wake wave without taking into account the influence of ion motion are derived, assuming a laser pulse of a given shape moving along the *OZ* axis at a velocity of $0 \leq v_g \leq c$, where c is the speed of light.

7.2.1. Equations in scalar form

We will assume that

- ions, due to the multiple excess of electrons by mass, are considered immobile;
- the solution is determined only by the following components of the vector and scalar functions: $\mathbf{v}_e = (V_r, V_z)$, $\mathbf{p}_e = (P_r, P_z)$, $\mathbf{E} = (E_r, E_z)$, n_e, γ, B_φ;
- there is no dependence in the specified functions on the variable φ, that is, $\dfrac{\partial}{\partial \varphi} = 0$.

Then, from system (7.1)–(7.8), written in cylindrical coordinates, we obtain non-trivial equations whose solutions have axial symmetry:

$$\frac{\partial n_e}{\partial t} + \frac{1}{r}\frac{\partial}{\partial r}(rn_eV_r) + \frac{\partial}{\partial z}(n_eV_z) = 0, \tag{7.9}$$

$$\frac{\partial P_r}{\partial t} = eE_r - m_e c^2 \frac{\partial \gamma}{\partial r}, \quad \frac{\partial P_z}{\partial t} = eE_z - m_e c^2 \frac{\partial \gamma}{\partial z}, \tag{7.10}$$

$$\gamma = \sqrt{1 + \frac{P_r^2 + P_z^2}{m_e^2 c^2} + \frac{|a|^2}{2}}, \tag{7.11}$$

$$V_r = \frac{P_r}{m_e \gamma}, \quad V_z = \frac{P_z}{m_e \gamma}, \tag{7.12}$$

$$\frac{\partial E_r}{\partial t} = -4\pi e n_e V_r - c \frac{\partial B_\varphi}{\partial z},$$

$$\frac{\partial E_z}{\partial t} = -4\pi e n_e V_z + c \frac{1}{r} \frac{\partial}{\partial r}(r B_\varphi), \tag{7.13}$$

$$B_\varphi = -\frac{c}{e} \left(\frac{\partial P_r}{\partial z} - \frac{\partial P_z}{\partial r} \right). \tag{7.14}$$

In deriving equations (7.9)–(7.14), new notation was not used, auxiliary transformations were not performed.

7.2.2. New coordinates and quasistatics

We assume that the wake wave is excited by a laser pulse moving along the *OZ* axis at a velocity of $0 \leq v_g \leq c$, and we introduce new independent variables characterizing the coordinate system associated with the pulse (the coordinate *r* remains the same):

$$\xi = z - v_g t, \quad t = \tau.$$

This corresponds to substitutions of derivatives in the system of equations (7.9)–(7.14):

$$\frac{\partial}{\partial t} \rightarrow \frac{\partial}{\partial \tau} - v_g \frac{\partial}{\partial \xi}, \quad \frac{\partial}{\partial z} \rightarrow \frac{\partial}{\partial \xi}.$$

In addition, we use the so-called quasistatic approximation, characterized by the ratio of the smallness of the derivatives

$$\frac{\partial}{\partial \tau} \ll v_g \frac{\partial}{\partial \xi}.$$

As a result, from equations (7.9)–(7.14) we get:

$$\frac{\partial}{\partial \xi}[n_e \ (V_z - v_g)] + \frac{1}{r} \frac{\partial}{\partial r} \ (rn_e V_r) = 0, \tag{7.15}$$

$$v_g \frac{\partial P_r}{\partial \xi} = -eE_r + m_e c^2 \frac{\partial \gamma}{\partial r}, \quad v_g \frac{\partial P_z}{\partial \xi} = -eE_z + m_e c^2 \frac{\partial \gamma}{\partial \xi}, \tag{7.16}$$

$$\gamma = \sqrt{1 + \frac{P_r^2 + P_z^2}{m_e^2 c^2} + \frac{|a|^2}{2}}, \tag{7.17}$$

$$V_r = \frac{P_r}{m_e \gamma}, \quad V_z = \frac{P_z}{m_e \gamma}, \tag{7.18}$$

$$v_g \frac{\partial E_r}{\partial \xi} = 4\pi e n_e V_r - \frac{c^2}{e} \frac{\partial}{\partial \xi}\left(\frac{\partial P_r}{\partial \xi} - \frac{\partial P_z}{\partial r}\right),$$

$$v_g \frac{\partial E_z}{\partial \xi} = 4\pi e n_e V_z + \frac{c^2}{e} \frac{1}{r} \frac{\partial}{\partial r}\left[r\left(\frac{\partial P_r}{\partial \xi} - \frac{\partial P_z}{\partial r}\right)\right], \tag{7.19}$$

$$B_\varphi = -\frac{c}{e} \left(\frac{\partial P_r}{\partial \xi} - \frac{\partial P_z}{\partial r}\right). \tag{7.20}$$

About the system (7.15)–(7.20), it should be noted that even if the appearance of the equations did not change compared to (7.9)–(7.14), the independent variables changed anyway: in the new system, all the required functions depend only on variables r and ξ.

7.2.3. Equations in dimensionless variables

We introduce dimensionless quantities

$$\rho = k_p r, \quad \eta = k_p \xi, \quad \beta_g = \frac{v_g}{c}, \quad q_r = \frac{P_r}{mc}, \quad q_z = \frac{P_z}{mc},$$

$$\varepsilon_r = \frac{eE_r}{mc^2 k_p}, \quad \varepsilon_z = \frac{eE_z}{mc^2 k_p}, \quad B = \frac{eB_\varphi}{mc^2 k_p}, \quad N = \frac{n_e}{n_0},$$

where $\omega_p = (4\pi e^2 n_0/m)^{1/2}$ is the plasma frequency, $k_p = \omega_p/c$, n_0 is the value of the unperturbed electron density.

In the new variables, after excluding V_r, V_z and B_φ, system (7.15)–(7.20) takes the form

$$\frac{\partial}{\partial \eta}\left[N\left(\frac{q_z}{\gamma} - \beta_g\right)\right] + \frac{1}{\rho}\frac{\partial}{\partial \rho}\left(\rho N \frac{q_r}{\gamma}\right) = 0, \tag{7.21}$$

$$\beta_g \frac{\partial q_r}{\partial \eta} + \varepsilon_r = \frac{\partial \gamma}{\partial \rho}, \quad \beta_g \frac{\partial q_r}{\partial \eta} + \varepsilon_r = \frac{\partial \gamma}{\partial \eta}, \tag{7.22}$$

$$\gamma = \sqrt{1 + q_r^2 + q_z^2 + \frac{|a|^2}{2}}, \tag{7.23}$$

$$\beta_g \frac{\partial \varepsilon_r}{\partial \eta} = N\frac{q_r}{\gamma} - \frac{\partial}{\partial \eta}\left(\frac{\partial q_r}{\partial \eta} - \frac{\partial q_z}{\partial \rho}\right),$$

$$\beta_g \frac{\partial \varepsilon_z}{\partial \eta} = N\frac{q_z}{\gamma} + \frac{1}{\rho}\frac{\partial}{\partial \rho}\left[\rho\left(\frac{\partial q_r}{\partial \eta} - \frac{\partial q_z}{\partial \rho}\right)\right]. \tag{7.24}$$

7.2.4. Equations in convenient variables

In the resulting system (7.21)–(7.24) there are six desired functions $(q_r,\ q_z,\ \varepsilon_r,\ \varepsilon_z,\ N,\ \gamma)$, depending on the independent variables ρ and η. It is very cumbersome and therefore difficult to study. It seems reasonable to reduce the number of required functions to four by entering three new values:

$$\psi = \beta_g q_z - \gamma, \quad \varphi = N/\gamma, \quad q = \beta_g q_r, \tag{7.25}$$

and leaving unchanged the function γ. Note that if the functions ψ, φ, q, γ are defined, then returning to the original unknowns is easy and unique.

Let us make a remark: since the notation φ, standard for the independent variable in cylindrical coordinates, is not used in the case under consideration, then φ in (7.25) denotes one of the new convenient functions depending on the variables ρ and η.

Express from the first equation (7.22) value

$$\varepsilon_r = \frac{\partial \gamma}{\partial \rho} - \beta_g \frac{\partial q_r}{\partial \eta}$$

and substitute it into the first equation (7.24). As a result, we get

$$\beta_g \frac{\partial}{\partial \eta}\left(\frac{\partial \gamma}{\partial \rho} - \beta_g \frac{\partial q_r}{\partial \eta}\right) = \frac{N}{\gamma} q_r - \frac{\partial}{\partial \eta}\left(\frac{\partial q_r}{\partial \eta} - \frac{\partial q_z}{\partial \rho}\right).$$

Multiply the obtained relation by β_g and exclude from it the value $\beta_g q_z = \psi + \gamma$; in the transition to new variables (7.25) we will have

$$\frac{\partial^2 \psi}{\partial \rho \partial \eta} + q\varphi - \varepsilon\left(\frac{\partial^2 q}{\partial \eta^2} - \frac{\partial^2 \gamma}{\partial \rho \partial \eta}\right) = 0. \tag{7.26}$$

Hereinafter, we use the notation $\varepsilon = 1-\beta_g^2$; taking into account $0 \le \beta_g \le 1$, we note that $\varepsilon \ge 0$.

Multiply equation (7.21) by β_g and eliminate $q\varphi$ from it using relation (7.26); will get

$$\frac{\partial}{\partial \eta}\left\{\Delta\psi - \varphi\psi + \varepsilon\left[\Delta\gamma - \varphi\gamma - \frac{1}{\rho}\frac{\partial}{\partial \rho}\left(\rho\frac{\partial q}{\partial \eta}\right)\right]\right\} = 0,$$

where $\Delta = \frac{1}{\rho}\frac{\partial}{\partial \rho}\left(\rho\frac{\partial}{\partial \rho}\right)$ is the radial part of the Laplace operator in cylindrical coordinates. Since the obtained expression is valid both inside the laser pulse and outside it, it is possible to determine the missing constants for integration from the values of the desired functions at rest. By definition, the limits of the functions ψ, φ, q, γ as $\rho \to \infty$, $\eta \to \infty$ are respectively -1, 1, 0, 1, hence we have

$$-\Delta\psi + \varphi\psi + 1 + \varepsilon\left[-\Delta\gamma + \varphi\gamma - 1 + \frac{1}{\rho}\frac{\partial}{\partial \rho}\left(\rho\frac{\partial q}{\partial \eta}\right)\right] = 0. \tag{7.27}$$

Express from the second equation (7.22) value

$$\varepsilon_z = \frac{\partial}{\partial \eta}(\gamma - \beta_g q_z) \equiv -\frac{\partial \psi}{\partial \eta}$$

and substitute it into the second equation (7.24), as a result we get

$$- \beta_g \frac{\partial^2 \psi}{\partial \eta^2} = \varphi q_z + \frac{1}{\rho} \frac{\partial}{\partial \rho} \left(\rho \frac{\partial q_r}{\partial \eta} \right) - \Delta q_z = 0.$$

The substitution $q_z = (\psi + \gamma)/\beta_g$ gives

$$- \beta_g \frac{\partial^2 \psi}{\partial \eta^2} + \frac{1}{\beta_g} [\Delta \psi - \varphi \psi + \Delta \gamma - \varphi \gamma] - \frac{1}{\rho} \frac{\partial}{\partial \rho} \left(\rho \frac{\partial q_r}{\partial \eta} \right) = 0.$$

Multiply the resulting equation by and add it up with (7.27), after reducing the similar and term by term reduction by the value $\varepsilon - 1 = - \beta_g^2$, we arrive at the equation

$$\frac{\partial^2 \psi}{\partial \eta^2} + \frac{1}{\rho} \frac{\partial}{\partial \rho} \left(\rho \frac{\partial q}{\partial \eta} \right) - \Delta \gamma + \varphi \gamma - 1 = 0. \tag{7.28}$$

Using (7.28), we can simplify equation (7.27) by reducing it to

$$- \Delta \psi + \varphi \psi + 1 - \varepsilon \frac{\partial^2 \psi}{\partial \eta^2} = 0. \tag{7.29}$$

Let us transform the last untouched equation from system (7.21)–(7.24) – equation (7.23). We will put both of them in a square and use substitutions

$$q_r = \frac{q}{\beta_g}, \quad q_z = \frac{\psi + \gamma}{\beta_g}, \quad \frac{1}{\beta_g^2} - 1 = \frac{\varepsilon}{1 - \varepsilon}.$$

As a result, we get

$$\psi^2 + 2 \psi \gamma + 1 + q^2 + \frac{|a|^2}{2} + \frac{\varepsilon}{1 - \varepsilon} \left[(\gamma + \psi)^2 + q^2 \right] = 0. \tag{7.30}$$

The transformation of equations for convenient desired variables is complete, new equations (7.26), (7.28)–(7.30) are obtained; we note in them reasonable values of the parameter ε: as a rule, the dimensionless value of the velocity of the laser pulse is in the range of $0.9 \leq \beta_g \leq 1$, which gives $0 \leq \varepsilon \leq 0.19$.

7.3. The basic formulation of the problem

The section formulates a nonlinear differential-algebraic problem, the solution of which has the property of breaking. The words 'hydrodynamic model of a wake wave' first of all means this formulation. In the future, all numerical algorithms will be designed to solve it. In addition, to verify the properties of numerical algorithms, a simpler (linearized) formulation is formulated, the solution of which does not possess the property of breaking.

7.3.1. Nonlinear statement

We consider an axially symmetric laser pulse propagating along the OZ axis with a group velocity v_g, which in a underdense plasma is close to the speed of light. In the basic formulation, we assume that $v_g = c$ and move to the coordinate system associated with the pulse. In it, the pulse, whose profile changes according to the Gaussian law, will be stationary. Neglecting the change in the pulse with time, i.e. for

$$a(\rho,\eta) = a_* \exp\left\{-\frac{\rho^2}{\rho_*^2} - \frac{\eta^2}{l_*^2}\right\}, \qquad (7.31)$$

from the system of equations (7.26), (7.28)–(7.30) with $\varepsilon = 0$ we get

$$\frac{\partial \psi}{\partial \eta} = \beta, \quad \frac{\partial \beta}{\partial \eta} + \frac{1}{\rho}\frac{\partial}{\partial \rho}\rho\frac{\partial q}{\partial \eta} - \Delta\gamma + \varphi\gamma - 1 = 0, \quad q\varphi + \frac{\partial \beta}{\partial \rho} = 0,$$

$$\varphi\psi - \Delta\psi + 1 = 0, \quad 2\gamma\psi + \psi^2 + q^2 + 1 + \frac{|a|^2}{2} = 0. \qquad (7.32)$$

In order to further the convenience of constructing numerical algorithms, a new variable $\beta = \partial\psi/\partial\eta$ has been artificially introduced here.

Consider the formulation of the initial and boundary conditions of the problem, based on the locality of the solution in the transverse direction and its axial symmetry.

For the physical variables q_r, q_z, ε_r, ε_z, N, γ, the equations have the dimensionless form (7.21)–(7.24). If we assume that the plasma occupies the whole space, then the domain of change of the independent variables ρ and η is unbounded: $\rho \in [0, \infty)$, $\eta \in$

$(-\infty, \infty)$. Moreover, on the axis of symmetry (for $\rho = 0$), due to the evenness/oddness of the desired functions, the boundary conditions for $\forall \eta \in (-\infty, \infty)$ are:

$$q_r = \varepsilon_r = 0, \quad \frac{\partial q_z}{\partial \rho} = \frac{\partial \varepsilon_z}{\partial \rho} = \frac{\partial N}{\partial \rho} = \frac{\partial \gamma}{\partial \rho} = 0. \tag{7.33}$$

In turn, at the radial periphery of the region ($\rho \to \infty$), the plasma is in an unperturbed state of rest, that is, for $\forall \eta \in (-\infty, \infty)$, the following holds:

$$q_r = \varepsilon_r = q_z = \varepsilon_z = 0, \quad N = \gamma = 1. \tag{7.34}$$

Since the laser pulse is stationary in the coordinate system (ρ, η), that is, the plasma 'runs' into a pulse, it is also in an unperturbed state of rest before the pulse ($\eta \to \infty$), therefore, for $\forall \rho \in [0, \infty)$ we have:

$$q_r = \varepsilon_r = q_z = \varepsilon_z = 0, \quad N = \gamma = 1,$$
$$\frac{\partial q_r}{\partial \eta} = \frac{\partial \varepsilon_r}{\partial \eta} = \frac{\partial q_z}{\partial \eta} = \frac{\partial \varepsilon_z}{\partial \eta} = \frac{\partial N}{\partial \eta} = \frac{\partial \gamma}{\partial \eta} = 0. \tag{7.35}$$

Conditions (7.33), (7.34) and formula (7.31) give rise to the possibility of transforming the solution domain from unbounded in the variables ρ and η to the bounded: $\rho \in [0, R]$, $\eta \in [Z_e, Z_s]$. The physical formulation of the problem assumes that the pulse cross section is substantially smaller than the transverse size of the region occupied by the plasma, that is, $\exp(-R^2/\rho_*^2) \ll 1$. For this, it suffices to set $R \approx 4.5\rho_*$. The longitudinal characteristics of a bounded region are chosen from other considerations: $\exp(-Z_s^2/l^2) \ll 1$ and the value of Z_e is determined only by research objectives and available computational resources. A typical value in the calculations is $Z_s \approx 3.5l_*$. Note that the variable η in the problem changes in the direction of decreasing, i.e., integration over it is carried out in the opposite direction.

Let us turn to the equations in convenient variables, i.e., to equations (7.32). The specified agreement of the geometric parameters of the problem actually means the modelling of the following situation. Before the pulse ($\eta \geq Z_s$) there is an unperturbed plasma, that is, when $0 \leq \rho \leq R$, the initial conditions are given

$$\psi(\rho, Z_s) = -1, \beta(\rho, Z_s) = 0, q(\rho, Z_s) = 0,$$
$$\varphi(\rho, Z_s) = 1, \gamma(\rho, Z_s) = 1.$$
(7.36)

Then (for $Z_e \leq \eta < Z_s$) the plasma changes its structure in accordance with the solution of system (7.32), however, in the cross section the pulse is localized so that the boundary values of the functions (for $\rho = R$, $\eta \geq Z_s$) can be considered unperturbed, i.e. relevant state of rest:

$$\psi(R, \eta) = -1, \beta(R, \eta) = 0, q(R, \eta) = 0, \varphi(R, \eta) = 1, \gamma(R, \eta) = 1;$$
(7.37)

on the axis (for $\rho = 0$, $\eta \geq Z_s$), due to the axial symmetry of the problem, the following conditions are set:

$$\frac{\partial \psi}{\partial \rho}(\rho = 0, \eta) = \frac{\partial \beta}{\partial \rho}(\rho = 0, \eta) = \frac{\partial \gamma}{\partial \rho}(\rho = 0, \eta)$$

$$= \frac{\partial \varphi}{\partial \rho}(\rho = 0, \eta) = q(\rho = 0, \eta) = 0.$$
(7.38)

The system of equations (7.31), (7.32) in the domain $\rho \in [0, R]$, $\eta \in [Z_e, Z_s]$, where $R = 4.5\rho_*$, $Z_s = 3.5l_*$, together with a set of initial and boundary conditions (7.36)–(7.38) will be referred to as the basic formulation of the problem for modelling wake plasma waves. Note that the stated formulation is a mixed initial–boundary value problem for a differential–algebraic nonlinear system of partial differential equations. The boundary conditions are defined at the boundaries of the variable ρ, and the initial conditions (as in the Cauchy problem) by the variable η with $\eta = Z_s$, since the variable η changes in the problem in the direction of decreasing. To summarize the above, the 'hydrodynamic model of the wake wave' is completely defined.

7.3.2. Linearized formulation

The nonlinearities in the basic formulation are quadratic. Therefore, it is easy to see that for small values of the amplitude of the laser pulse a_* in formula (7.31), the solutions of system (7.32) will slightly differ from their background values corresponding to the unperturbed plasma state (7.37). We obtain linear equations for small perturbations of solutions (7.32).

Let's make in the system of equations (7.32) the replacement of the desired variables of the form

$$\tilde{\psi} = \psi + 1, \ \tilde{\varphi} = \varphi - 1, \ \tilde{\gamma} = \gamma - 1, \ \tilde{\beta} = \beta, \ \tilde{q} = q, \tag{7.39}$$

after which we omit the quadratic terms in it with respect to the functions labelled 'tilde'. We assume them in order substantially smaller than the linear terms. As a result, we get:

$$\frac{\partial \tilde{\beta}}{\partial \eta} + \frac{1}{\rho} \frac{\partial}{\partial \rho} \rho \frac{\partial \tilde{q}}{\partial \eta} - \Delta \tilde{\gamma} + \tilde{\varphi} + \tilde{\gamma} = 0, \tag{7.40}$$

$$\tilde{q} = -\frac{\partial \tilde{\beta}}{\partial \rho}, \quad \tilde{\varphi} = \tilde{\psi} - \Delta \tilde{\psi}, \quad \tilde{\beta} = \frac{\partial \tilde{\psi}}{\partial \eta}, \quad \tilde{\gamma} = \frac{|a|^2}{4}. \tag{7.41}$$

Substituting explicit expressions for $\tilde{q}, \tilde{\varphi}, \tilde{\beta}$ in (7.40) gives:

$$(E - \Delta) \left(\frac{\partial^2 \tilde{\psi}}{\partial \eta^2} + \tilde{\psi} + \tilde{\gamma} \right) = 0 \tag{7.42}$$

where E is the identity operator.

The boundary conditions for the functions $\tilde{\psi}$ and $\tilde{\gamma}$, which follow from (7.37), (7.38), generate from (7.42) the model equation for the function $\tilde{\psi}(\rho, \eta)$:

$$\frac{\partial^2 \tilde{\psi}}{\partial \eta^2} + \tilde{\psi} + \frac{|a|^2}{4} = 0. \tag{7.43}$$

It should be noted that in this equation the variable ρ is present only as a parameter participating in the formula for the laser pulse (7.31).

The final linearized formulation for the basic problem can be formulated as follows: in the domain $\rho \in [0, R]$, $\eta \in [Z_e, Z_s]$, define the functions β and ψ, which are solutions to the Cauchy problem:

$$\frac{\partial \tilde{\psi}}{\partial \eta} = \tilde{\beta}, \ \frac{\partial \tilde{\beta}}{\partial \eta} + \tilde{\psi} + \frac{|a|^2}{4} = 0, \quad \tilde{\psi}(\rho, Z_s) = \tilde{\beta}(\rho, Z_s) = 0. \tag{7.44}$$

If necessary, the remaining functions $\tilde{q}, \tilde{\varphi}, \tilde{\gamma}$ can be uniquely determined from relations (7.41).

An important feature of the transition to the linearized formulation is the change in the type of equations: from a differential–algebraic system of partial differential equations, a system of ordinary differential equations with a parametric dependence on the spatial variable is obtained. This imposes additional restrictions on the research methods of discrete analogs of the basic formulation.

We make a special remark that the algorithms designed for a nonlinear basic formulation must withstand all discrete transformations similar to the differential transformations that were considered in this section and lead to a discrete analogue of problem (7.44).

7.4. 'Slow' pulse

The section considers the situation when, in contrast to the basic formulation, the pulse velocity is less than the speed of light, that is, $\varepsilon > 0$. As shown below, this leads to an exponentially unstable problem even for the linearized formulation. A combined numerical–asymptotic method is proposed as an approach to the solution, which partially alleviates the problem.

7.4.1. Linearized equations

Recall that equations (7.26), (7.28)–(7.30) describe the case of an arbitrary pulse velocity. As a rule, the dimensionless value of the velocitu of a laser pulse is in the range of $0.9 \leq \beta_g \leq 1$, which gives $0 \leq \varepsilon \leq 0.19$.

Let us reduce the equations to a form convenient for analysis. We normalize the independent spatial variable $\tau = \eta/\sqrt{1-\varepsilon}$ and introduce the notation $q = p/\sqrt{1-\varepsilon}$, $\psi = -\sigma$. Then the equations (7.26), (7.28)–(7.30) take the following form:

$$1 + \sigma^2 - 2\sigma\gamma + p^2 + \frac{|a|^2}{2} + \frac{\varepsilon}{1-\varepsilon}(\gamma - \sigma)^2 = 0,$$

$$\frac{\partial^2 \sigma}{\partial \rho \partial \tau} - p\varphi + \varepsilon\left(p\varphi + \frac{\partial^2 p}{\partial \tau^2} - \frac{\partial^2 \gamma}{\partial \rho \partial \tau}\right) = 0,$$

$$\Delta_\perp \sigma - \varphi\sigma + 1 + \frac{\varepsilon}{1-\varepsilon}\frac{\partial^2 \sigma}{\partial \tau^2} = 0,$$

$$\frac{\partial^2 \sigma}{\partial \tau^2} - \frac{1}{\rho}\frac{\partial}{\partial \rho}\left(\rho\frac{\partial p}{\partial \tau}\right) + \Delta_\perp \gamma - \gamma\varphi + 1 + \frac{\varepsilon}{1-\varepsilon}\frac{\partial^2 \sigma}{\partial \tau^2} = 0.$$

$$(7.45)$$

In equations (7.45), as in the basic formulation of the problem, the plasma motion is initialized by the pulse envelope, which, in neglect of the pulse variation with time, is described by the formula

$$a(\rho,\eta) = a_* \exp\left\{-\frac{\rho^2}{\rho_*^2} - \frac{\eta^2}{l_*^2}\right\}.$$

We will assume that for small values of the amplitude of the laser pulse a_*, the solutions of system (7.45) will slightly differ from their background values corresponding to the unperturbed plasma state (7.37). We obtain in the usual way linear equations for small perturbations of solutions.

Let's make in the system of equations (7.45) the replacement of the desired variables

$$\sigma = 1 + \tilde\psi, \quad \gamma = 1 + \tilde\gamma, \quad \varphi = 1 + \tilde\varphi, \quad p = \tilde q$$

and neglect the quadratic terms due to their smallness; will get

$$\tilde\gamma = \frac{|a|^2}{4},$$

$$\frac{\partial^2 \tilde\psi}{\partial\rho\partial\tau} - \tilde q = \varepsilon\left(\frac{\partial^2 \tilde\gamma}{\partial\rho\partial\tau} - \tilde q - \frac{\partial^2 \tilde q}{\partial\tau^2}\right),$$

$$\Delta_\perp\tilde\psi - \tilde\varphi - \tilde\psi + \frac{\varepsilon}{1-\varepsilon}\frac{\partial^2 \tilde\psi}{\partial\tau^2} = 0, \qquad (7.46)$$

$$\frac{\partial^2 \tilde\psi}{\partial\tau^2} - \frac{1}{\rho}\frac{\partial}{\partial\rho}\left(\rho\frac{\partial\tilde q}{\partial\tau}\right) + \Delta_\perp\tilde\gamma - \tilde\varphi - \tilde\gamma + \frac{\varepsilon}{1-\varepsilon}\frac{\partial^2 \tilde\psi}{\partial\tau^2} = 0.$$

Let us study the properties of the linear system of equations (7.46) with a view to its suitability for numerical solution, i.e., first of all, with respect to stability.

7.4.2. Auxiliary Cauchy problem

In the domain $\Omega = \{(\rho, \tau): R \geq \rho \geq 0, \tau \geq 0\}$, we consider the auxiliary Cauchy problem for a linear system of equations

$$\frac{\partial^2 \psi}{\partial\rho\partial\tau} - q = \varepsilon\left(\frac{\partial^2 \gamma}{\partial\rho\partial\tau} - q - \frac{\partial^2 q}{\partial\tau^2}\right),$$

$$\frac{\partial^2 \psi}{\partial\tau^2} - \frac{1}{\rho}\frac{\partial}{\partial\rho}\left(\rho\frac{\partial q}{\partial\tau}\right) - \Delta_\perp(\psi - \gamma) + \psi - \gamma = 0, \qquad (7.47)$$

with respect to the unknown functions $\psi(\rho, \tau)$ and $q(\rho, \tau)$ with given boundary

$$\left.\frac{\partial \psi(\rho,\tau)}{\partial \rho}\right|_{\rho=0} = q(0,\tau) = \psi(R,\tau) = q(R,\tau) = 0 \quad \forall \tau \geq 0, \qquad (7.48)$$

and initial conditions for $\tau = 0$, $R \geq \rho \geq 0$:

$$\psi(\rho,0) = A(\rho), \quad \left.\frac{\partial \psi(\rho,\tau)}{\partial \tau}\right|_{\tau=0} = B(\rho), \qquad (7.49)$$

$$q(\rho,0) = C(\rho), \quad \left.\frac{\partial q(\rho,\tau)}{\partial \tau}\right|_{\tau=0} = D(\rho). \qquad (7.50)$$

Here it is assumed that the initial functions are sufficiently smooth and satisfy the boundary conditions, $\Delta_\perp = \dfrac{1}{\rho}\dfrac{\partial}{\partial \rho}\left(\rho\dfrac{\partial}{\partial \rho}\right)$ – is the radial part of the Laplace operator, $1 > \varepsilon \geq 0$ is the parameter of the problem. In addition, the system (7.47) has inhomogeneity, since the function $\gamma(\rho, \tau)$ is defined in a special way:

$$\gamma(\rho,\tau) = \frac{a_*^2}{4}\exp\left\{-\frac{2\rho^2}{\rho_*^2} - \frac{2(\tau - \tau_c)^2}{\tau_*^2}\right\}, \qquad (7.51)$$

where a_*, ρ_*, τ_c, τ_* are some pulse parameters, moreover, $\rho_* \ll R$, $\tau_* \ll \tau_c$.

We first consider the general solution of the homogeneous (that is, for $\gamma(\rho, \tau) \equiv 0$) problem (7.47)–(7.50). For this we present the solution in the form

$$\psi(\rho,\tau) = \sum_{k=1}^{\infty}\psi_k(\tau)Y_k(\rho), \quad q(\rho,\tau) = \sum_{k=1}^{\infty}q_k(\tau)Z_k(\rho), \qquad (7.52)$$

where

$$Y_k(\rho) = \frac{\sqrt{2}}{R\,|J_1(\lambda_k)|}J_0\left(\lambda_k\frac{\rho}{R}\right), Z_k(\rho) = \frac{\sqrt{2}}{R\,|J_1(\lambda_k)|}J_1\left(\lambda_k\frac{\rho}{R}\right)$$

– systems of functions satisfying the boundary conditions (7.48); $J_0(t)$, $J_1(t)$ are the Bessel functions; λ_k, $k = 1, 2 \ldots$ are the zeros of $J_0(t)$.

Completeness and orthonormality of each of the systems $\{Y_k\}_{k=1}^{\infty}, \{Z_k\}_{k=1}^{\infty}$ (when using the scalar product $(u,v) = \int_0^R \rho u v d\rho$ and the norm $\|u\| = (u, u)^{1/2}$) generate from (7.47) for each fixed k equation with respect to the expansion coefficients (7.52):

$$v_k \psi_k' + q_k = \varepsilon(q_k'' + q_k), \quad \psi_k'' - v_k q_k' + (v_k^2 + 1)\psi_k = 0, \qquad (7.53)$$

where $v_k = \lambda_k/R$, and the prime means differentiation with respect to the variable τ. The characteristic equation of this system is

$$(\lambda^2 + 1)\left[\varepsilon\lambda^2 - (1-\varepsilon)(1+v_k^2)\right] = 0,$$

where the expressions for the general solution of equations (7.53) follow:

$$\psi_k(\tau) = \sum_{j=1}^{4} H_j^{(1)} g_j(\tau), \quad q_k(\tau) = \sum_{j=1}^{4} H_j^{(2)} g_j(\tau),$$

where $\{g_j(\tau)\}_{j=1}^{4}$ is the fundamental system:

$$g_1(\tau) = \sin(\tau), g_2(\tau) = \cos(\tau), g_3(\tau) = \exp(E_k\tau), g_4(\tau) = \exp(-E_k\tau),$$

with the coefficient in the last two functions, $E_k = [(1-\varepsilon)(1+v_k^2)/\varepsilon]^{1/2}$, and $H_j^{(i)}$, $i = 1, 2$, are determined from the corresponding expansions of the initial functions (7.49), (7.50) in the series of the form (7.52). From the above calculations follows

Statement 7.4.1. *Suppose that in (7.47) ε is a fixed number from (0, 1) and two positive numbers are given: g is an arbitrarily small number, and G is an arbitrarily large number. Then for any given $\tau > 0$ there are sufficiently smooth initial functions in (7.49), (7.50) satisfying the condition* $\max \{\|A\|, \|B\|, \|C\|, \|D\|\} \le g$, *and such that to solve the homogeneous system (7.47)–(7.50) the inequality* $\max \{\|\psi\|, \|q\|\} \ge G$. *will be true.*

Statement 7.4.1 means that for $\varepsilon \ne 0$, the solution of problem (7.47)–(7.50) is exponentially unstable, as in the well-known example of Hadamard, which demonstrates the instability of the classical Cauchy problem for the Laplace equation [87] such solution can not be found numerically in principle.. However, the problem under consideration has a specific property that makes it possible to avoid direct integration of the system (7.47) in order to find an approximate solution.

Consider the stable special case of the problem (7.47)–(7.50) with $\varepsilon = 0$, which requires additional specification of the conditions (7.48), (7.49) and generates from (7.47) the decomposing system of equations

$$\frac{\partial^2 \psi}{\partial \tau^2} + \psi - \gamma = 0, \quad q = \frac{\partial^2 \psi}{\partial \rho \partial \tau}. \tag{7.54}$$

There is valid.

Lemma 7.4.1. *Let* $\psi_0(\rho, \tau)$, $q_0(\rho, \tau)$ *be a sufficiently smooth solution of the inhomogeneous system* (7.54), (7.49) *with given* γ (ρ, τ), *then the function* q_0 *satisfies the equality in* Ω

$$\frac{\partial^2 \gamma}{\partial \rho \partial \tau} - q_0 - \frac{\partial^2 q_0}{\partial \tau^2} = 0.$$

Based on the lemma it is established:

Statement 7.4.2. *Let the function* $\gamma(\rho, \tau)$ *of the form* (7.51) *be fixed with* $\tau_c \geq \tau_*$, *find the solution of the problem* (7.54), (7.49) ψ_0, q_0 *and define the initial functions in* (7.50) *as follows:*

$$C(\rho) = q_0(\rho, 0), \quad D(\rho) = \frac{\partial q_0(\rho, \tau)}{\partial \tau}\bigg|_{\tau=0}.$$

Then the solution of the problem (7.47)–(7.50) ψ_ε, q_ε, *corresponding to the value* $\varepsilon \neq 0$, *identically coincides with* ψ_0, q_0, *i.e.*

$$\psi_\varepsilon(\rho, \tau) = \psi_0(\rho, \tau), \quad q_\varepsilon(\rho, \tau) = q_0(\rho, \tau) \quad \forall (\rho, \tau) \in \Omega.$$

Statement 7.4.2 is useful to provide two comments. First, the independence of the solution from the parameter ε concerns only the system of equations (7.47), and not the original more general system of equations describing laser–plasma interactions. The fact is that already after the transition from the general case to the linear formulation (for $a_* \ll 1$), a variable of the form $\tau = \eta / \sqrt{1 - \varepsilon}$ was replaced, leading to the analyzed form of equations (7.47). In other words, the dependence on ε is present in a hidden form (in the form of the dependence of τ on ε and, respectively, in the inhomogeneity $\gamma(\rho, \tau)$) even in the solution ψ_0, q_0, if the latter is considered in the original independent variables ρ and t. Second: the procedure for selecting additional initial data may seem artificial, since it is necessary to replace the finding of the solution ψ_ε, q_ε by $-\psi_0$, q_0, and not vice versa. However, from a practical point of view, this moment is completely unimportant, since the physical process itself starts from a state of complete rest (all functions and derivatives are assumed to be zero at the initial moment of time) due to the condition $\tau_c \gg \tau_*$.

7.4.3. Numerical–asymptotic method

The auxiliary Cauchy problem considered in the previous section is the 'content' part of the linearized formulation in the case of a 'slow' pulse (7.46), since equations (7.47) follow from equations (7.46) as a result of eliminating the function $\tilde{\varphi}$ and further lowering the symbol ~.

From the analysis in the previous section, we have two important consequences:

1) as a result of omitting the the term

$$\varepsilon\left(\frac{\partial^2 \tilde{\gamma}}{\partial \rho \partial \tau} - p - \frac{\partial^2 p}{\partial \tau^2}\right)$$

in the first equation (7.47), based on the transformations from section 7.3.2, we obtain a stable method for solving a linear problem;

2) the corresponding equations describing the first approximation for direct expansion in a small parameter ε of the solution of the original nonlinear problem (7.45) do not contain resonant terms (at least, in an explicit form).

Based on the above consequences, we can formulate a numerical-asymptotic method for simulating wake waves generated by a 'slow' pulse. We describe it in series.

Consider the direct ε–decomposition

$$f = f_0 + \varepsilon f_1 + \varepsilon^2 f_2 + \dots.$$

for all unknown functions from equations (7.45).

Then, to determine the zero approximation to the solution, we obtain from (7.45) a nonlinear system

$$1 + \psi_0^2 - 2\psi_0\gamma_0 + p_0^2 + \frac{|a|^2}{2} = 0,$$

$$\frac{\partial^2 \psi_0}{\partial\rho\partial\tau} - p_0\varphi_0 = 0,$$

$$\Delta_\perp \psi_0 - \varphi_0\psi_0 + 1 = 0,$$

$$\frac{\partial^2 \psi_0}{\partial\tau^2} - \frac{1}{\rho}\frac{\partial}{\partial\rho}\left(\rho\frac{\partial p_0}{\partial\tau}\right) + \Delta_\perp\gamma_0 - \gamma_0\varphi_0 + 1 = 0.$$

(7.55)

This system of equations coincides with the equations of the basic formulation (7.31), (7.32). Therefore, adding to them the set of initial and boundary conditions (7.36)–(7.38), one can obtain a zero approximation to the desired solution by any of the methods considered for the basic formulation of the methods.

Having a zero approximation available, it is easy to obtain also from (7.45) a system of linear equations for determining the first approximation:

$$2\psi_0\psi_1 - 2\psi_0\gamma_1 - 2\psi_1\gamma_0 + 2p_0p_1 + (\gamma_0 - \psi_0)^2 = 0,$$

$$\frac{\partial^2 \psi_1}{\partial\rho\partial\tau} - p_0\varphi_1 - p_1\varphi_0 + \left(p_0\varphi_0 + \frac{\partial^2 p_0}{\partial\tau^2} - \frac{\partial^2 \gamma_0}{\partial\rho\partial\tau}\right) = 0,$$

$$\Delta_\perp\psi_1 - \varphi_0\psi_1 - \varphi_1\psi_0 + \frac{\partial^2 \psi_0}{\partial\tau^2} = 0,$$

$$\frac{\partial^2 \psi_1}{\partial\tau^2} - \frac{1}{\rho}\frac{\partial}{\partial\rho}\left(\rho\frac{\partial p_1}{\partial\tau}\right) + \Delta_\perp\gamma_1 - \gamma_0\varphi_1 - \gamma_1\varphi_0 + \frac{\partial^2 \psi_0}{\partial\tau^2} = 0.$$

(7.56)

As noted above, these linear equations do not contain resonant terms, so they can be used for long time intervals. In addition, all «unstable» terms in them are computed explicitly on the basis of the already known zero approximation.

Preliminary calculations demonstrated the following influence of the 'slow' pulse:

1) the wake wave dynamics shifts closer to the pulse center, i.e., fully corresponds to the transformation of the longitudinal coordinate $\eta = \tau\sqrt{1-\varepsilon}$;

2) changes in the regular part of the wave are almost imperceptible;

3) in the irregular part of the wave (in the vicinity of the breakingzone) the maximum of the electron density is less by about 5%, and the

minimum is more by about 28%, i.e. a «smoothing» of the most sensitive function $N(\rho, \eta)$ occurs.

More detailed systematic studies related to the excitation of wake waves by a 'slow' pulse have not yet been carried out.

7.5. Bibliography and comments

The article [43] seems to be extremely useful for a deeper understanding of the need to study the properties of wake waves excited in plasma by laser pulses, .

The idca of using lasers to accelerate electrons in a plasma was launched in 1979 [166]. For short laser pulses, the first analytical studies were published in the late 1980s [44, 162]. In fact, the laser acceleration of electrons in plasma is very close to the so-called collective method of electron acceleration, which was developed for many years at the Kharkov Institute of Physics and Technology back in Soviet times, when scientific research was conducted on a broad front, and their results were a special national pride.

At present, considerable experimental and theoretical material (in France, USA, Japan, England) [133] has been accumulated, sufficient for the design and construction of a laser electron accelerator up to an energy of more than 1000 MeV. Several such projects are close to implementation.

The main method for modelling wake waves in a plasma, as already noted in the preface, are PiC models. A fairly large number of books are devoted to their description (see [50, 115, 139] and the literature cited there); It is useful to add to this list a review article [167] and a domestic view on the history of the approach in the book [23].

In this monograph, the hydrodynamic model of the wake wave, developed under the direction of L.M. Gorbunov at the Lebedev Physical Institute (FIAN, Moscow). Its appearance was preceded by a series of works on the theory of three-dimensional waves excited by a laser pulse; among them should be noted [9, 10, 113]. An analytical theory of quasi-plane wake waves excited in an underdense plasma by intense laser pulses was constructed in [10], and a linear theory of the structure of the wake wave in plasma channels was used in [113]. In fact, these works gave a start to the numerical simulationn of plasma wakefield waves, which is referred to in the book. Similar ideas were simultaneously developed in the works of P.A. Mora and T.M. Antonsen; their useful work [153] should be mentioned, based on the kinetic model of the propagation of a powerful short laser pulse.

Pioneering studies related to the structure of the plasma wake wave, which allowed for the possibility of breaking after several periods, were carried out in [10].

A significant difference from other wake wave models is the use of the relation

$$\mathbf{B}(\mathbf{x},t)+\frac{c}{e}\ \mathrm{curl}\ \mathbf{p}_e(\mathbf{x},t)=0,$$

arising from the initial condition (1.14) of a special form.

For stationary nonrelativistic equations, this relation is well known (see, for example, [37, 78]) and is associated, first of all, with irrotational plasma motion. For non-stationary relativistic equations, it was derived in [3, 4] and called the 'law of conservation of a generalized curl'.

The transformation of equations describing the dynamics of the wake wave, using variables convenient for numerical solution, is given in [109].

The system of equations in the form (7.45) for simulating a wake wave excited by a laser pulse moving at an arbitrary velocity $0 \le v_g \le c$ appears to be presented for the first time.

Results demonstrating the exponential instability of equations in the presence of a 'slow' pulse were obtained in [91, 92].

When a wake wave is excited by a laser pulse, it is generally required to solve the equations for the slowly varying amplitude of the high-frequency laser field (the so-called envelope). In order not to clutter the presentation, the subject related to the propagation of a pulse in a substance is not considered in the book. This subject was studied in a separate series of works where the following aspects were: transient nonlinear waves with ponderomotive self-focusing of radiation in a plasma [8], dynamics of self-focusing of a wave beam in a plasma [12], thermal self-focusing in a plasma [47], etc. original numerical algorithms based on a combination of known methods for calculating the nonlinear Schrödinger equations and thermal conductivity, as well as taking into account kinetically consistent difference schemes and quasi-gasdynamic equations. We present the essence of these works in somewhat more detail.

The propagation of an intense beam of electromagnetic radiation in a plasma can be accompanied by its self-focusing, as a result of which the transverse size of the beam decreases and the intensity of the radiation increases [105]. This phenomenon arises due to a change of the dielectric constant of the plasma under the action of electromagnetic radiation.

The usual theory of stationary or quasi-stationary self-focusing of electromagnetic beams is based on the assumption that a nonlinear response is established in a time shorter than the time of intensity change (see, for example, [73, 105]). However, many types of non-linearities in a plasma (for example, thermal, ionization, ponderomotive) are determined by relatively slow processes and the time of their establishment can be much longer than the time of intensity change in the beam. In this case, the nature of the self-focusing process is largely determined by the dynamics of the formation of the nonlinear response.

With the plasma parameters considered in [8], the main mechanism determining the nonlinear response is the redistribution of plasma density under the action of the averaged ponderomotive force (high-frequency pressure force). In work [114], it was shown that in a transient process with ponderomotive self-focusing, a nonlinear density and radiation intensity wave is excited, propagating along the beam axis from the boundary into the plasma with a velocity substantially exceeding the speed of sound. In [114], the assumption of small perturbations of plasma density was used. In the framework of such an approximation, stationary self-focusing is described by a nonlinear Schrödinger equation with a cubic nonlinearity where the axially symmetric solution, as is well known (see, for example, [57, 105]), has a singularity (focus). To eliminate this feature, nonlinear dissipation of the electromagnetic field was introduced in Ref. [114], which actually corresponds to multiphoton absorption in gases [73], but does not have any physical justification for a fully ionized plasma.

In plasma, the physical reason for the absence of singularity in the steady state is the nonlinearity of the density perturbations created by ponderomotive forces. To describe the non-stationary process of radiation self-action, it is necessary to use nonlinear hydrodynamic equations for both the plasma density and its velocity in an electromagnetic field [42]. It is on the basis of such equations that the dynamics of self-focusing of an axially symmetric beam were studied in [8, 12]. The results of similar quasistationary calculations have been published in a number of papers (see, for example, [127, 152]).

Unlike work [114], according to which a transient nonlinear wave propagates only from the plasma boundary and to the point where a stationary focus is established, it is shown in [8] that the transient wave continuously goes into the plasma and is a characteristic feature of the self-focusing process in environments without nonlinear dissipation and with a relatively long nonlinear response time. Behind the nonlinear wave, a steady state is established, which is a sequence of maxima of the radiation intensity on the beam axis. An increase in the time during which the radiation intensity at the plasma boundary reaches its stationary value leads to an increase in the distance from the boundary at which the transition nonlinear wave begins to be excited.

The joint solution of nonstationary equations describing plasma dynamics and the electric field intensity of radiation required the development of a new numerical algorithm, which was based on the construction of a kinetically consistent difference scheme for the Euler equations (see [55, 56, 97] and the literature cited there). The calculation results showed that such an approach makes it possible to obtain reliable results in a wide range of plasma parameters, despite the fact that the analogue of the gas-dynamic equation of state has, in general, a non-standard character.

The solution of a nonlinearized system of hydrodynamic equations for plasma density and its velocity together with a nonlinear equation for the envelope avoided the singularity problem arising in a stationary state in a theory based on linearized acoustic equations [114]. It was shown that in a transient process a nonlinear plasma density wave and radiation intensity are excited, propagating along the beam axis from the boundary into the plasma depth with a velocity substantially exceeding the speed of sound.

The model studied in [8, 114, 127, 152] assumes that the electron plasma temperature is constant. This assumption is justified in rare and hot plasmas, where the electron mean free path is much larger than the transverse size of the beam.

In [47], the process of self-focusing was investigated for another limiting case – a small length of the free path of electrons, when the heating of plasma particles is the main one (thermal self-focusing). A complete system of hydrodynamic equations was used to describe plasma [30]. It is shown that in this case the steady state is a sequence of intensity maxima on the beam axis. However, in the process of reaching this steady state, a nonlinear wave does not arise, although the transient process is not monotonous.

The joint solution of nonstationary equations describing plasma dynamics and the electric field intensity of radiation required the development of a new numerical algorithm based on a combination of the known methods for calculating nonlinear Schrödinger equations and thermal conductivity [21, 26, 63, 84] with non-standard kinetics consistent difference schemes and quasi-gasdynamic equations [55, 56, 97].

We note once again that the simulation of self-focusing effects of a laser pulse is not considered in the monograph. However, in chapter 9, for completeness of presentation and taking into account the relevance of modelling laser-plasma interactions, the algorithms for calculating the envelope with reference to the wake wave excitation are described.

Numerical algorithms for the basic problem

This chapter designs and analyzes methods for the approximate solution of the basic formulation of a nonlinear differential algebraic problem, based on various constructive ideas. Numerical experiments demonstrate the behaviour of the solution of the hydrodynamic model of the wake wave.

8.1. Difference method I

8.1.1. Construction of a difference scheme

We construct in the domain $\Omega = \{(\rho, \eta): 0 \leq \rho \leq R, Z_e \leq \eta \leq Z_s\}$ a uniform grid Ω_h in both variables with steps h_r and h_z so that

$$\rho_m = mh_r, \quad 0 \leqslant m \leqslant M, \quad R = h_r M; \quad \eta_j = Z_e + jh_z,$$
$$0 \leqslant j \leqslant N_z, \quad Z_s = Z_e + h_z N_z,$$

and proceed to the description of the difference scheme and the algorithm for its implementation for the basic nonlinear problem formulated above, that is, equations (7.31), (7.32) together with a set of initial and boundary conditions (7.36)–(7.38). In this case, for grid functions in the node (ρ_m, η_j), we use notation of the form f_m^j, omitting some indices where this should not cause any confusion.

On the grid Ω_h we define four of the five desired functions: β, ψ, φ, γ. We define the function q on a slightly different grid, namely: in the nodes of the main grid Ω_h that are shifted by 0.5 h_r and associated with the variable ρ. This will be indicated by fractional indices; for example, $q_{m+1/2}^{j}$ means the value of the grid function at the node $(\rho_m + 0.5\,h_r, \eta_j = Z_e + jh_z)$.

We write the approximation of the equation

$$\frac{\partial \beta}{\partial \eta} + \frac{1}{\rho}\frac{\partial}{\partial \rho}\rho\frac{\partial q}{\partial \eta} - \Delta\gamma + \varphi\gamma - 1 = 0$$

in the internal nodes of the grid Ω_h, i.e. when $j = N_z - 1, N_z - 2, \ldots, 0; m = = 1, 2, \ldots, M - 1$:

$$\frac{\beta_m^{j+1} - \beta_m^{j}}{h_z} + \frac{1}{\rho_m h_r}\left(\rho_{m+\frac{1}{2}}\frac{q_{m+\frac{1}{2}}^{j+1} - q_{m+\frac{1}{2}}^{j}}{h_z} - \rho_{m-\frac{1}{2}}\frac{q_{m-\frac{1}{2}}^{j+1} - q_{m-\frac{1}{2}}^{j}}{h_z}\right) - \qquad (8.1)$$

$$- \Delta^h\gamma_m^j + \varphi_m^j\gamma_m^j - 1 = 0.$$

Here for the radial part of the Laplace operator (using cylindrical coordinates)

$$\Delta = \frac{1}{r}\frac{\partial}{\partial r}\left(r\frac{\partial}{\partial r}\right)$$

in the node ρ_m, $m \geq 1$, the usual approximation is used [22]:

$$\Delta^h f_m = \frac{1}{\rho_m}\frac{1}{h_r}\left(\rho_{m+\frac{1}{2}}\frac{f_{m+1} - f_m}{h_r} - \rho_{m-\frac{1}{2}}\frac{f_m - f_{m-1}}{h_r}\right).$$

On the axis of symmetry, i.e., for $m = 0, j = N_z - 1, N_z - 2, \ldots, 0$, the approximation of the equation will have a slightly different form:

$$\frac{\beta_0^{j+1} - \beta_0^{j}}{h_z} + \frac{2}{h_r}\left(\frac{q_{+\frac{1}{2}}^{j+1} - q_{+\frac{1}{2}}^{j}}{h_z} - \frac{q_{-\frac{1}{2}}^{j+1} - q_{-\frac{1}{2}}^{j}}{h_z}\right) - \qquad (8.2)$$

$$- \Delta\gamma_0^j + \varphi_0^j\gamma_0^j - 1 = 0,$$

where

$$\Delta^h f_0 = \frac{4}{h_r^2}(f_1 - f_0)$$

– the difference analog of the Laplace operator on the symmetry axis with allowance for the condition of the evenness of the function $f(\rho)$ in the neighbourhood of the straight line $\rho = 0$ [22].

In a similar way, we first write the approximation of the equation

$$q\varphi + \frac{\partial \beta}{\partial \rho} = 0$$

in the internal nodes of the grid, that is, with $j = N_z - 1, N_z - 2, ...,$ 0; $m = = 1, 2, ..., M - 1$:

$$q^j_{m-\frac{1}{2}} \frac{\varphi^j_m + \varphi^j_{m-1}}{2} + \frac{\beta^j_m - \beta^j_{m-1}}{h_r} = 0, \qquad (8.3)$$

and then on the axis of symmetry ($m = 0, j = N_z - 1, N_z - 2, ..., 0$):

$$q^j_{-\frac{1}{2}} \frac{\varphi^j_0 + \varphi^j_1}{2} + \frac{\beta^j_0 - \beta^j_1}{h_r} = 0. \qquad (8.4)$$

From relation (8.3) with $m = 1$ and the formula (8.4) it follows that

$$\frac{1}{2}(q^j_{-\frac{1}{2}} + q^j_{\frac{1}{2}}) = 0, \quad j = N_z - 1, N_z - 2, ..., 0,$$

which is a second order approximation for the condition $q(\rho = 0, \eta) = 0$ on the axis of symmetry of the domain Ω.

It should also be noted here that the approximation of this differential equation is carried out at the nodes of the auxiliary grid, not the main one. Therefore, the product of the functions $q\varphi$ is replaced by the value at the node of the function q multiplied by the half-sum of the values of the function φ taken at the neighbouring nodes with respect to the variable ρ. This technique allows, in the nodes of the auxiliary grid for smooth functions, to obtain in the equation under consideration an error of the second order approximation with respect to the value of h_r.

The remaining equations

$$\varphi\psi - \Delta\psi + 1 = 0, \quad 2\gamma\psi + \psi^2 + q^2 + 1 + \frac{|a|^2}{2} = 0, \quad \frac{\partial\psi}{\partial\eta} - \beta = 0$$

replace with the following discrete analogues in the grid nodes Ω_h:

$$\varphi_m^j \psi_m^j - \Delta^h \psi_m^j + 1 = 0,$$

$$2\gamma_m^j \psi_m^j + (\psi_m^j)^2 + q_{m-\frac{1}{2}}^j q_{m+\frac{1}{2}}^j + 1 + \frac{\left|a_m^j\right|^2}{2} = 0, \qquad (8.5)$$

$$\frac{\psi_m^{j+1} - \psi_m^j}{h_z} - \beta_m^j = 0,$$

both inside the Ω_h region and on the axis of symmetry – with $j = N_z - 1$, $N_z - 2$, ..., 0; $m = 0, 1, ..., M - 1$. In this case, it should be noted that for a smooth q function, the approximation of q^2 by the product $q_{m-\frac{1}{2}} q_{m+\frac{1}{2}}$ leads to the second-order smallness error with respect to the discretization parameter h_r.

To close the reduced difference equations, one should add the missing initial and boundary conditions, which in this case are conditions of 'rest', that is, they characterize the unperturbed plasma. When $j = N_z$; $m = 0, 1, ..., M$, and also for $m = M$; $j = N_z, N_z - 1, ..., 0$ we have:

$$\psi_m^j = -1, \quad \beta_m^j = q_{m-\frac{1}{2}}^j = 0, \quad \varphi_m^j = \gamma_m^j = 1. \qquad (8.6)$$

Here, attention should be paid to the formal approximation of the first order of the boundary condition for the function q – 'shifting' of zero value by 0.5 h_r. For $m = M$, this does not affect the quality of the resulting solution due to the exponentially small values of the function itself, and for $m = 0$, zero values are taken only for $j = N_z$, that is, in the region related to the initially unperturbed plasma. The above means that in this case the 'boundary' condition does not reduce the rate of convergence of the solution of the difference scheme as a whole.

The proposed difference scheme has in the internal nodes of the grid the second order of approximation with respect to the steps in ρ and the first in η; stability and, therefore, convergence with the same orders for the linearized problem can be obtained using the transformations described in the next section.

8.1.2. Study of schemes in variations

For the difference scheme (8.1)–(8.6), we write the system of equations for small variations (changes) of the unknown grid functions.

First, we note that if the function a_m^j, which initializes the wake wave, is identically zero, then the solution of the considered difference scheme is just the background values of the desired functions (8.6):

$$\psi_m^j = -1, \quad \beta_m^j = q_{m-\frac{1}{2}}^j = 0, \quad \varphi_m^j = \gamma_m^j = 1 \quad \forall m \in [0, M], \forall j \in [0, N].$$

Next, we represent the desired solution in the form

$$\psi_m^j = -1 + \tilde{\psi}_m^j, \quad \beta_m^j = 0 + \tilde{\beta}_m^j, \quad q_{m-\frac{1}{2}}^j = 0 + \tilde{q}_{m-\frac{1}{2}}^j,$$

$$\varphi_m^j = 1 + \tilde{\varphi}_m^j, \quad \gamma_m^j = 1 + \tilde{\gamma}_m^j, \tag{8.7}$$

where the 'tilde' mark indicates variations of the grid functions.

Now, assuming that the laser pulse is sufficiently small $|a_m^j| \ll 1$ we substitute the expressions (8.7) into the difference scheme (8.1)–(8.6) and discard the terms of the second order in the relations obtained with respect to the 'tilde' values. As a result, we obtain a linear system of difference equations with constant coefficients, written down with respect to small variations.

Relations (8.1), (8.2) are converted to the form

$$\frac{\tilde{\beta}_m^{j+1} - \tilde{\beta}_m^j}{h_z} + \frac{1}{\rho_m h_r}\left(\rho_{m+\frac{1}{2}} \frac{\tilde{q}_{m+\frac{1}{2}}^{j+1} - \tilde{q}_{m+\frac{1}{2}}^j}{h_z} - \rho_{m-\frac{1}{2}} \frac{\tilde{q}_{m-\frac{1}{2}}^{j+1} - \tilde{q}_{m-\frac{1}{2}}^j}{h_z} \right) -$$

$$- \Delta^h \tilde{\gamma}_m^j + \tilde{\varphi}_m^j + \tilde{\gamma}_m^j = 0, \tag{8.8}$$

$$\frac{\tilde{\beta}_0^{j+1} - \tilde{\beta}_0^j}{h_z} + \frac{2}{h_r}\left(\frac{\tilde{q}_{+\frac{1}{2}}^{j+1} - \tilde{q}_{+\frac{1}{2}}^j}{h_z} - \frac{\tilde{q}_{-\frac{1}{2}}^{j+1} - \tilde{q}_{-\frac{1}{2}}^j}{h_z} \right) - \Delta \tilde{\gamma}_0^j + \tilde{\varphi}_0^j + \tilde{\gamma}_0^j = 0.$$

Equations (8.3), (8.4) will change in the same way:

$$\tilde{q}_{m-\frac{1}{2}}^j + \frac{\tilde{\beta}_m^j - \tilde{\beta}_{m-1}^j}{h_r} = 0, \quad \tilde{q}_{-\frac{1}{2}}^j + \frac{\tilde{\beta}_0^j - \tilde{\beta}_1^j}{h_r} = 0, \tag{8.9}$$

from which the approximation of the boundary condition for the function q on the axis of symmetry of the domain also follows:

$$\frac{1}{2}(\tilde{q}^{j}_{-\frac{1}{2}} + \tilde{q}^{j}_{\frac{1}{2}}) = 0, \quad j = N_z - 1, N_z - 2, \ldots, 0.$$

Difference equations (8.5) written relative to the variations will take the form

$$\tilde{\varphi}^{j}_{m} = \tilde{\psi}^{j}_{m} - \Delta^{h}\tilde{\psi}^{j}_{m}, \quad \tilde{\gamma}^{j}_{m} = \frac{|a^{j}_{m}|^{2}}{4}, \quad \frac{\tilde{\psi}^{j+1}_{m} - \tilde{\psi}^{j}_{m}}{h_z} - \tilde{\beta}^{j}_{m} = 0. \qquad (8.10)$$

Finally, the initial and boundary conditions (8.6) for variations of the desired functions will become uniform and the same:

$$\tilde{\psi}^{j}_{m} = \tilde{\beta}^{j}_{m} = \tilde{q}^{j}_{m-\frac{1}{2}} = \tilde{\varphi}^{j}_{m} = \tilde{\gamma}^{j}_{m} = 0. \qquad (8.11)$$

We substitute explicit expressions for the functions $\tilde{q}^{j}_{m-\frac{1}{2}}$ of (8.9) and $\tilde{\varphi}^{j}_{m}$ from (8.10) into relations (8.8), as a result for $m = 0, 1, \ldots,$ $M - 1, j = N_z - 1, N_z - 2, \ldots 0$ get

$$(I - \Delta^{h})\,(\frac{\tilde{\beta}^{j+1}_{m} - \tilde{\beta}^{j}_{m}}{h_z} + \tilde{\psi}^{j}_{m} + \tilde{\gamma}^{j}_{m}) = 0, \qquad (8.12)$$

where I denotes the identity operator.

Taking into account the discrete boundary conditions for the functions $\tilde{\beta}^{j}_{m}$, $\tilde{\psi}^{j}_{m}$, $\tilde{\gamma}^{j}_{m}$ (evenness conditions in the vicinity of the axis of symmetry and the 'rest' conditions on the periphery of the domain), the difference equation follows from (8.12)

$$\frac{\tilde{\beta}^{j+1}_{m} - \tilde{\beta}^{j}_{m}}{h_z} + \tilde{\psi}^{j}_{m} + \tilde{\gamma}^{j}_{m} = 0. \qquad (8.13)$$

Now, if from (8.10) to the resulting equation (8.13) add the equations

$$\tilde{\gamma}^{j}_{m} = \frac{|a^{j}_{m}|^{2}}{4}, \quad \frac{\tilde{\psi}^{j+1}_{m} - \tilde{\psi}^{j}_{m}}{h_z} - \tilde{\beta}^{j}_{m} = 0, \qquad (8.14)$$

then, together with the initial conditions from (8.11), we have the Cauchy difference problem with respect to unknown variations $\tilde{\beta}^{j}_{m}$, $\tilde{\psi}^{j}_{m}$ for a given variation $\tilde{\gamma}^{j}_{m}$.

It is easy to see that difference equations (8.13), (8.14) approximate the system of differential equations

$$\frac{\partial \tilde{\beta}}{\partial \eta} + \tilde{\psi} = -\frac{|a(\rho,\eta)|^2}{4}, \quad \frac{\partial \tilde{\psi}}{\partial \eta} - \tilde{\beta} = 0$$

with the first order with respect to the discretization parameter h_z.

Considering that integration over the variable η is carried out in the direction of its decrease, that is, a decrease in the index j in the difference equations, then from (8.13), (8.14) the characteristic equation follows

$$(1+h_z^2)\mu^2 - 2\mu + 1 = 0,$$

having complex conjugate roots $\mu_{1,2} = \dfrac{1 \pm ih_z}{1+h_z^2}$, each of which is strictly less than one in absolute value. Hence, by virtue of the theorem of A.F. Filippov (see, for example, [22]), we obtain the convergence of an approximate solution of a linear differential equation to an exact first-order solution with respect to the discretization parameter h_z.

Based on the above considerations, we finally conclude that the difference scheme, written down with respect to small variations of the desired grid functions, is absolutely stable, and its solution converges to the solution of the linearized problem (7.44) with first order with respect to h_z. In this case, the remaining unknown variations $\tilde{q}^j_{m-\frac{1}{2}}$ and $\tilde{\varphi}^j_m$ are determined by explicit second-order approximation formulas with respect to the parameter h_r.

8.1.3. The algorithm for implementing the difference scheme I

Recall that the wake wave model is considered in the approximation of a given pulse, that is, the function $a^j_m = a(\rho_m, \eta_j)$ is considered known at any node of the grid Ω_h by formula (7.31). In this case, the algorithm for implementing the reduced nonlinear difference scheme is as follows: system (8.1)–(8.6) integrates over η in the direction of decreasing the index j, i.e., successively for each $j = N_z - 1$, $N_z - 2$, ..., 0 the three-point vector system is solved by the Newton method, implemented by a matrix (5×5) tridiagonal (Thomas) algorithm. We clarify what has been said with the help of useful formalization.

Imagine the desired solution in the form (8.7):

$$\psi_m^j = -1 + \tilde{\psi}_m^j, \quad \beta_m^j = 0 + \tilde{\beta}_m^j, \quad q_{m-\frac{1}{2}}^j = 0 + \tilde{q}_{m-\frac{1}{2}}^j,$$

$$\varphi_m^j = 1 + \tilde{\varphi}_m^j, \quad \gamma_m^j = 1 + \tilde{\gamma}_m^j,$$

where the 'tilde' mark, as before, marks the variations of the grid functions. However, here the smallness of the laser pulse is not assumed; therefore, the quadratic terms remain in the equations and make them nonlinear.

Let us fix in system (8.1)–(8.6) some value of the index j. Since the integration over the variable η is carried out in the direction of its decrease, the values of the functions marked with the index $j + 1$ are considered already known (given or already calculated). Determine the vector of unknowns

$$Y \equiv Y_m = (\tilde{\beta}_m^j, \tilde{q}_{m-1/2}^j, \tilde{\varphi}_m^j, \tilde{\gamma}_m^j, \tilde{\psi}_m^j)^T, \quad m = 0, 1, \ldots, M$$

(in fact, unknowns form a $5 \times (M + 1)$ matrix and rewrite equations (8.1)–(8.6) as a nonlinear algebraic system $F(Y) = 0$. To solve it, we use the classical Newton method [22]:

$$F'(Y^n)(Y^{n+1} - Y^n) + F(Y^n) = 0,$$
$$Y^0 = (\tilde{\beta}_m^{j+1}, \tilde{q}_{m-1/2}^{j+1}, \tilde{\varphi}_m^{j+1}, \tilde{\gamma}_m^{j+1}, \tilde{\psi}_m^{j+1})^T, \quad m = 0, 1, \ldots, M. \tag{8.15}$$

In this case, the auxiliary systems of linear equations at each Newtonian iteration $F'(Y^n)X = D$, where $X = Y^{n+1} - Y^n$, $D = -F(Y^n)$ have the following form:

$$C_0 X_0 - B_0 X_1 = D_0, \qquad\qquad m = 0,$$
$$-A_0 X_{m-1} + C_m X_m - B_m X_{m+1} = D_m, \quad 1 \leqslant m \leqslant M - 1, \tag{8.16}$$
$$X_M = 0, \qquad\qquad m = M,$$

where X_m and D_m are vectors of dimension 5, A_m, B_m, C_m are matrices of dimension 5×5.

To solve systems of this type (three-point vector equations), the matrix Thomas (diagonal) method [83] is very convenient. In it, the running coefficients (matrices and vectors) are first defined recursively in the direction of increasing the index $m = 1, 2, \ldots, M - 1$:

$$\alpha_{m+1} = (C_m - A_m \, \alpha_m)^{-1} B_m, \qquad \alpha_1 = C_0^{-1} B_0,$$
$$\beta_{m+1} = (C_m - A_m \, \alpha_m)^{-1} (D_m + A_m \, \beta_m), \qquad \beta_1 = C_0^{-1} D_0, \tag{8.17}$$

and then the components of the vector of unknowns are determined in the direction of decreasing index $m = M - 1, M - 2, ..., 0$:

$$X_m = \alpha_{m+1} X_{m+1} + \beta_{m+1}, \quad X_M = \mathbf{0}. \tag{8.18}$$

The correctness and stability of the method (8.17), (8.18) for solving the system (8.16) can be easily checked for the linearized formulation of relatively small variations considered in the previous section. The fact is that the matrix Thomas method presented here is actually equivalent to solving equation (8.12) with 'rest' conditions and even/odd conditions in the vicinity of the axis of symmetry for variations of the desired functions.

The convergence of the Newton method was fixed by the maximum in $m = 0, 1, ..., M - 1$ and all equations with a residual not exceeding $\delta = 10^{-4}$. As a rule, in the calculations this corresponded to no more than three iterations.

8.2. Difference method II

The difference method I considered in the previous section has a significant defect, namely: the corresponding difference scheme has a first-order approximation error with respect to the parameter h_z. In particular, in calculations, when the wake wave propagates over a large number of periods, such an error may be unacceptably large. In addition, the difference method I is very laborious: the matrix (5×5)-Thomas (tridiagnonal) algorithtm on each Newtonian iteration of each step in the variable η can lead to unjustified utilization of computational resources.

In this section we will try to overcome both difficulties at the same time, that is, to construct a more accurate and 'faster' (computationally less time-consuming), but, still, difference method.

8.2.1. Construction of a difference scheme

The discrete analog of the domain $\Omega = \{(\rho, \eta): 0 \le \rho \le R, Z_e \le \eta \le Z_s\}$ has not changed: Ω_h is a uniform grid in both variables with steps h_r and h_z

$$\rho_m = mh_r, \quad 0 \leqslant m \leqslant M, R = h_r M;$$
$$\eta_j = Z_e + jh_z, \quad 0 \leqslant j \leqslant N_z, \quad Z_s = Z_e + h_z N_z.$$

As before, we will use for the grid functions in the node (ρ_m, η_j) notation of the form f_m^j, omitting some indices where this should not cause confusion.

On the grid Ω_h, we define four of the five desired functions: β, ψ, φ, γ. We define the function q on a slightly different grid, namely: in the nodes of the main grid Ω_h that are shifted by 0.5 h_r and associated with the variable ρ. This will be indicated by fractional indices; for example, $q_{m+1/2}^j$ means the value of the grid function at the node $(\rho_m + 0.5\ h_r, \eta_j = Z_e + jh_z)$.

We immediately concentrate on the main thing: the main difference is concentrated in the approximation of equations

$$\varphi\psi - \Delta\psi + 1 = 0, \quad 2\gamma\psi + \psi^2 + q^2 + 1 + \frac{|a|^2}{2} = 0, \quad \frac{\partial\psi}{\partial\eta} - \beta = 0.$$

In the previous paragraph, their discrete analogs had the form (8.5):

$$\varphi_m^j \psi_m^j - \Delta^h \psi_m^j + 1 = 0, \quad 2\gamma_m^j \psi_m^j + (\psi_m^j)^2 + q_{m-\frac{1}{2}}^j q_{m+\frac{1}{2}}^j + 1 + \frac{|a_m^j|^2}{2} = 0,$$

$$\frac{\psi_m^{j+1} - \psi_m^j}{h_z} - \beta_m^j = 0.$$

In the proposed modification, the approximation scheme is somewhat different:

$$\varphi_m^j \psi_m^j - \Delta^h \psi_m^j + 1 = 0,$$

$$2\gamma_m^j \psi_m^j + (\psi_m^j)^2 + \left(\frac{q_{m-\frac{1}{2}}^j + q_{m+\frac{1}{2}}^j}{2}\right)^2 + 1 + \frac{|a_m^j|^2}{2} = 0, \tag{8.19}$$

$$\frac{\psi_m^{j+1} - \psi_m^j}{h_z} - \beta_m^{j+1} = 0.$$

Note that the changes occurred in the second equation (q^2 turned into a half-sum of neighbouring values in a square) and in the third equation (the superscript for the variable β changed). The remaining difference equations (8.1)–(8.4) and, accordingly, the initial and boundary conditions (8.6) have not undergone any changes.

To study the approximation, stability, and convergence of the new difference scheme, we turn to the model (linearized) formulation.

8.2.2. Study of schemes in variations

For the difference scheme (8.1)–(8.4), (8.6), (8.19), we write the system of equations for small variations (changes) of unknown grid functions.

As before, we represent the desired solution as a shift relative to the background:

$$\psi_m^j = -1 + \tilde{\psi}_m^j, \quad \beta_m^j = 0 + \tilde{\beta}_m^j, \quad q_{m-\frac{1}{2}}^j = 0 + \tilde{q}_{m-\frac{1}{2}}^j,$$

$$\varphi_m^j = 1 + \tilde{\varphi}_m^j, \quad \gamma_m^j = 1 + \tilde{\gamma}_m^j, \tag{8.20}$$

where the 'tilde' mark indicates the variations of the grid functions, and under the assumption that the laser pulse is sufficiently small $|a_m^j| \ll 1$ substitute the expressions (8.20) into the considered difference scheme. After discarding the second-order terms of smallness with respect to the 'tilde' in the relations obtained, we obtain a linear system of difference equations with constant coefficients, written down with respect to small variations.

Since only part of the difference equations has changed, we will use the results of the previous section and add to them the linearization of the differing equations (8.19):

$$\tilde{\varphi}_m^j = \tilde{\psi}_m^j - \Delta^h \tilde{\psi}_m^j, \quad \tilde{\gamma}_m^j = \frac{|a_m^j|^2}{4}, \quad \frac{\tilde{\psi}_m^{j+1} - \tilde{\psi}_m^j}{h_z} - \tilde{\beta}_m^{j+1} = 0. \tag{8.21}$$

In the linear case, the difference from relations (8.10) became even smaller; it exists only in the approximation of the equation $\dfrac{\partial \psi}{\partial \eta} - \beta = 0$.

Recall the missing difference equation (8.13):

$$\frac{\tilde{\beta}_m^{j+1} - \tilde{\beta}_m^j}{h_z} + \tilde{\psi}_m^j + \tilde{\gamma}_m^j = 0.$$

Taking into account the fact that integration over the variable η is carried out in the direction of its decrease, that is, a decrease in the index j in the difference equations, the characteristic equation follows from (8.13), (8.21)

$$\mu^2 - (2 - h_z^2)\mu + 1 = 0,$$

whose roots at $0 < h_z < 2$ are complex-conjugate and equal in modulus to one. Hence, by virtue of the theorem proposed by A.F. Filippov (see, for example, [22]), with the above condition, we obtain the convergence of an approximate solution of a linear differential equation to an exact solution with a second order with respect to the discretization parameter h_z.

Based on the above considerations, we finally conclude that the difference scheme, written down with respect to small variations of the desired grid functions, is absolutely stable, and its solution converges to solving the linearized problem (7.44) with second order with respect to h_z. In this case, the remaining unknown variations $\tilde{q}^j_{m-\frac{1}{2}}$ and $\tilde{\varphi}^j_m$ are determined by explicit second-order approximation formulas with respect to the parameter h_r.

8.2.3. Algorithm for the implementation of difference scheme II

Recall that the wake wave model is considered in the approximation of a given pulse, that is, the function $a^j_m = a(\rho_m, \eta_j)$ is considered known at any node of the grid Ω_h by the formula (7.31). In this case, the algorithm for implementing the reduced nonlinear difference scheme is as follows: system (8.1)–(8.4), (8.19) with the initial and boundary conditions (8.6) is integrated by η in the direction of decreasing the index j. We clarify what has been said with the help of useful formalization.

Imagine the desired solution in the form (8.7):

$$\psi^j_m = -1 + \tilde{\psi}^j_m, \quad \beta^j_m = 0 + \tilde{\beta}^j_m, \quad q^j_{m-\frac{1}{2}} = 0 + \tilde{q}^j_{m-\frac{1}{2}},$$

$$\varphi^j_m = 1 + \tilde{\varphi}^j_m, \quad \gamma^j_m = 1 + \tilde{\gamma}^j_m,$$

where the 'tilde' mark, as before, marks the variations of the grid functions. However, here the smallness of the laser pulse is not assumed; therefore, the quadratic terms remain in the equations and make them nonlinear.

Let us fix in system (8.1)–(8.4), (8.19) some value of the index j. Since the integration over the variable η is in the direction of its decrease, the values of the functions marked with the index $j + 1$ are considered already known (given or calculated).

First, from (8.19) we define the variation $\tilde{\psi}_m^j$:

$$\tilde{\psi}_m^j = \tilde{\psi}_m^{j+1} - h_z \tilde{\beta}_m^{j+1}, \quad m = 0,1,\ldots,M-1, \tag{8.22}$$

then, using the already found $\tilde{\psi}_m^j$, we calculate the variation $\tilde{\varphi}_m^j$ using the explicit formula:

$$\tilde{\varphi}_m^j = \frac{\tilde{\psi}_m^j - \Delta^h \tilde{\psi}_m^j}{1 - \tilde{\psi}_m^j}, \quad m = 0,1,\ldots,M-1. \tag{8.23}$$

For the remaining unknowns, the three-point vector system is solved by the Newton method implemented by the matrix (3×3)-Thomas algorithm. Determine the vector of the unknowns

$$Y \equiv Y_m = (\tilde{\beta}_m^j, \tilde{q}_{m-1/2}^j, \tilde{\gamma}_m^j)^T, \quad m = 0,1,\ldots,M$$

(in fact, unknowns form a $3 \times (M + 1)$ matrix) and rewrite the remaining equations (8.1)–(8.4), (8.19) with the initial and boundary conditions (8.11) with respect to the vector Y in the form of a nonlinear algebraic system $F(Y) = 0$. To solve it, we use the classical Newton method [22]:

$$F'(Y^n)(Y^{n+1} - Y^n) + F(Y^n) = 0,$$
$$Y^0 = (\tilde{\beta}_m^{j+1}, \tilde{q}_{m-1/2}^{j+1}, \tilde{\gamma}_m^{j+1})^T, \quad m = 0,1,\ldots,M. \tag{8.24}$$

In this case, the auxiliary systems of linear equations at each Newtonian iteration $F'(Y^n)X = D$, where $X = Y^{n+1} - Y^n$, $D = -F(Y^n)$ have the form (8.16), where the dimension 5 should be replaced by the dimension 3. For solving three-point vector equations, the matrix tridiagonal algorithm method [83] is convenient. It first recurrently determines the coefficients of the Thomas algotirhm (matrices and vectors) to increase the index $m = 1, 2, \ldots, M - 1$ by formulas (8.17), and then determine the components of the vector of unknowns in the direction of decreasing index $m = M - 1$, $M - 2, \ldots, 0$ by the formulas (8.18).

The correctness and stability of the method (8.17), (8.18) for solving the system (8.16) can be easily checked for the linearized formulation of relatively small variations considered in the previous section. The fact is that the matrix tridiagonal (Thomas) algorithm method presented here is actually equivalent to solving equation (8.12) with 'rest' conditions and even/odd conditions in the vicinity of the axis of symmetry for variations of the desired functions.

The convergence of the Newton method was fixed by the maximum in $m = 0, 1, ..., M - 1$ and all equations with a residual not exceeding $\delta = 10^{-4}$. As a rule, in the calculations this corresponded to no more than three iterations.

Note the improvements in the difference method I carried out in this section. First, the accuracy of the difference scheme with respect to the variable η increased by an order of magnitude: from $O(h_z)$ to $O(h_z^2)$. Secondly, when implemented by lowering the dimension of the matrices from 5 to 3, the computational costs in the matrix Thomas method were reduced, asymptotically – by $(5/3)^3 \approx 5$ times. This means that a more accurate and 'faster' (computationally more effective), but, as before, difference method is built in the section.

8.3. Difference method III (Linearization method)

Above, implicit finite–difference methods for calculating wake waves were considered. This section discusses an explicit scheme for the finite difference method for solving the same problem, which is simpler to implement and more economical in terms of computational costs.

8.3.1. Setting the task in a convenient form

Let us rewrite the basic formulation of the problem for modelling wake plasma waves (equations (7.31), (7.32) together with a set of initial and boundary conditions (7.36)–(7.38)) in a form more convenient for presentation of the algorithm.

Consider a system of partial differential algebraic equations

$$\mathbf{F}(\mathbf{x}) = \mathbf{y}, \tag{8.25}$$

where $\mathbf{x} \equiv \mathbf{x}(\rho, \eta) = (q, \varphi, \psi, \gamma)^{\mathrm{T}}$ is the vector of the unknowns, $\mathbf{y} = (0, 0, 0, |a|^2/4)^{\mathrm{T}}$ is the given right-hand side, $\mathbf{F} = (f_1, f_2, f_3, f_4)^{\mathrm{T}}$ is a nonlinear operator:

$$f_1 = q + \frac{\partial^2 \psi}{\partial \rho \partial \eta} + q\varphi, \quad f_2 = \varphi - \psi + \Delta\psi - \varphi\psi,$$

$$f_3 = \frac{\partial^2 \psi}{\partial \eta^2} + \frac{1}{\rho}\frac{\partial}{\partial \rho}\rho\frac{\partial q}{\partial \eta} - \Delta\gamma + \varphi + \gamma + \varphi\gamma, \quad f_4 = \gamma - \gamma\psi + \frac{1}{2}\left[\psi^2 + q^2\right].$$

Here, as before, the notation $\Delta = \frac{1}{\rho}\frac{\partial}{\partial \rho}\left(\rho\frac{\partial}{\partial \rho}\right)$ is used for the radial part of the Laplace operator.

The system (8.25) describes in a dimensionless form the propagation of an axially symmetric wake wave in a cold, ideal, relativistic electron liquid (plasma), initiated by a laser pulse with a given amplitude (the so-called envelope)

$$a(\rho,\eta) = a_* \exp\left\{-\frac{\rho^2}{\rho_*^2} - \frac{\eta^2}{l_*^2}\right\}. \tag{8.26}$$

The pulse parameters a_*, ρ_*, l_* actually determine the geometrical dimensions of the domain in which the solution of the initial–boundary value problem for equations (8.25) is considered. These equations are written in the coordinates associated with the pulse; therefore, integration over the variable η is conducted in the direction of its decrease; the centre of the pulse is considered fixed and based on the origin of coordinates ($\rho = \eta = 0$).

Equations (8.25)–(8.26) are considered in the domain

$$\Omega = \left\{(\rho,\eta) : 0 \leqslant \rho \leqslant R, \eta \leqslant Z_s\right\},$$

where the values of R and Z_s determine the boundary of the unperturbed plasma. For example, if the wake wave is determined by a function of the form (8.26), it suffices to set $R = 4 \div 4.5\rho_*$, $Z_s = 3.5 \div 4.5l_*$. The initial equations (8.25) contain partial derivatives, therefore they must be supplied with the initial and boundary conditions:

when $\eta = Z_s$, the state of rest is set, i.e., unperturbed plasma:

$$q = \varphi = \psi = \frac{\partial \psi}{\partial \eta} = \gamma = 0 \quad \forall \, 0 \leqslant \rho \leqslant R; \tag{8.27}$$

when $\rho = 0$, the conditions of axial symmetry are given

$$q = \lim_{\rho \to 0}\rho\frac{\partial \varphi}{\partial \rho} = \lim_{\rho \to 0}\rho\frac{\partial \psi}{\partial \rho} = \lim_{\rho \to 0}\rho\frac{\partial \gamma}{\partial \rho} = 0 \quad \forall \, \eta \leq Z_s; \tag{8.28}$$

for $\rho = R$, an unperturbed plasma is given:

$$q = \varphi = \psi = \gamma = 0 \quad \forall \, \eta \leqslant Z_s. \tag{8.29}$$

Problem (8.25)–(8.29) has a well-known specificity: an axially symmetric wake wave, excited by a laser pulse of the form (8.26), sooner or later breaks. This effect is associated with the intersection of the trajectories of electrons that transfer energy to the wave. The system (8.25) is written in the Euler variables, therefore, the breaking is observed as going back to infinity (singularity) of the electron density perturbation

$$N(\rho, \eta) = 1 + \varphi + \gamma + \varphi \gamma. \tag{8.30}$$

8.3.2. Preliminary transformations

Let us analyze the properties of the operator **F** in the system (8.25). We first divide the linear and nonlinear components so that (8.25) is transformed to

$$\mathbf{F}'\mathbf{x} + \mathbf{F}''(\mathbf{x}) = \mathbf{y}, \quad \mathbf{F} \equiv \mathbf{F}' + \mathbf{F}''.$$

Here $\mathbf{F}'' = (f_1'', f_2'', f_3'', f_4'')^T$ is the nonlinear part:

$$f_1'' = q\varphi, \quad f_2'' = -\varphi\psi, \quad f_3'' = \varphi\gamma, \quad f_4'' = -\gamma\psi + \frac{1}{2}\left[\psi^2 + q^2\right].$$

Then we will carry out the first factorization of the linear part. Direct check leads to:

Statement 8.3.1. *The* **F**' *operator can represented as*

$$\mathbf{F}' = LR, \tag{8.31}$$

where L is a lower triangular and R is an upper triangular operator:

$$L = \begin{pmatrix} I & 0 & 0 & 0 \\ 0 & I & 0 & 0 \\ \dfrac{1}{\rho}\dfrac{\partial}{\partial \rho}\rho\dfrac{\partial}{\partial \eta} & I & I & 0 \\ 0 & 0 & 0 & I \end{pmatrix}, \quad R = \begin{pmatrix} I & 0 & \dfrac{\partial^2}{\partial \rho \partial \eta} & 0 \\ 0 & I & \Delta - I & 0 \\ 0 & 0 & (I-\Delta)\left(\dfrac{\partial^2}{\partial \eta^2}+I\right) & I-\Delta \\ 0 & 0 & 0 & I \end{pmatrix}.$$

Statement 8.3.1 implies the possibility of making equations (8.25) more convenient

$$Rx = L^{-1}(y - F''(x)), \tag{8.32}$$

which follows from the invertibility of a nondegenerate matrix operator L by explicit formulas.

We will carry out the second factorization of the linear part of the problem, that is, the operator R. Also, we establish directly by checking:

Statement 8.3.2. *The operator R in* (8.32) *can be represented as*

$$R = D\tilde{R}, \tag{8.33}$$

where D is diagonal, and $\tilde{R}$ is the upper-triangular operators:

$$D = \begin{pmatrix} I & 0 & 0 & 0 \\ 0 & I & 0 & 0 \\ 0 & 0 & I-\Delta & 0 \\ 0 & 0 & 0 & I \end{pmatrix}, \quad \tilde{R} = \begin{pmatrix} I & 0 & \dfrac{\partial^2}{\partial\rho\,\partial\eta} & 0 \\ 0 & I & \Delta - I & 0 \\ 0 & 0 & \dfrac{\partial^2}{\partial\eta^2}+I & I \\ 0 & 0 & 0 & I \end{pmatrix}.$$

It should be noted that the inversion of the operator D essentially relies on the solution of the boundary value problem of the form

$$(I-\Delta)u \equiv -\frac{1}{\rho}\frac{\partial}{\partial\rho}\left(\rho\frac{\partial u}{\partial\rho}\right) + u = g(\rho),$$

$$\lim_{\rho\to 0}\rho\frac{\partial u}{\partial\rho} = 0, \quad u(R) = 0, \tag{8.34}$$

with some given function $g(\rho)$, and only the trivial solution corresponds to the zero right side in (8.34).

At this, the preliminary transformations end, since factorization (8.33) allows us to give the desired form to equations (8.25)

$$\tilde{R}x = D^{-1}L^{-1}(y - F''(x)) \equiv y - G''(x), \tag{8.35}$$

where the nonlinear operator $\mathbf{G}^n = (g_1, g_2, g_3, g_4)^T$ has the form:

$$g_1 = q\varphi, \quad g_2 = -\varphi\psi, \quad g_3 = (I - \Delta)^{-1}\left(\varphi\psi - \frac{1}{\rho}\frac{\partial}{\partial\rho}\rho\frac{\partial}{\partial\eta}q\varphi + \varphi\gamma\right),$$

$$g_4 = -\gamma\psi + \frac{1}{2}\left[\psi^2 + q^2\right].$$

8.3.3. Difference method III in the linear case

We present an algorithm for solving equations (8.35) under the assumption that the nonlinear terms are small. We note that in this case, the solution of problem (8.34) is not required; therefore, we will take into account the dependence of the desired functions on ρ as a dependence on some external parameter.

We introduce a uniform grid with respect to the variable η such that $\eta^j = Z_s + j\tau, j = 0, -1, -2, \ldots$. Recall that the variable η changes in the basic problem in the direction of decreasing.

The solution of the equation $\tilde{R}\mathbf{x} = \mathbf{y}$ is implemented in three stages:

1) the calculation of $\gamma^j(\rho) = |a^j(\rho)|^2/4$;

2) integration of the Cauchy problem for the linear equation $\psi'' + \psi + \gamma = 0$, which should be carried out according to the scheme

$$\frac{\psi^{j+1}(\rho) - \psi^j(\rho)}{\tau} - \beta^{j+1}(\rho) = 0,$$

$$\frac{\beta^{j+1}(\rho) - \beta^j(\rho)}{\tau} + \psi^j(\rho) + \gamma^j(\rho) = 0,$$

(8.36)

based on reduction to two first-order equations, since this significantly reduces the effect of computational error [22];

3) calculation, if necessary (for example, for periodic monitoring of electron density) using explicit formulas

$$q^j(\rho) = -\frac{\partial\beta^j(\rho)}{\partial\rho}, \quad \varphi^j(\rho) = \psi^j(\rho) - \Delta\psi^j(\rho).$$

Scheme (8.36) is equivalent to the standard scheme of the second order of accuracy with respect to τ, and the specifics of the implementation of the algorithm as a whole should include the ability to calculate $\psi^j(\rho)$ at each step, and only then sequentially $\gamma^j(\rho)$ and $\beta^j(\rho)$.

To complete the description of the linear scheme, one should determine discretization by variable ρ: the main grid is $\rho_m = mh$, $0 \le m \le M$, $Mh = R$; all functions except q are defined at the nodes of the main grid, and the function q is defined at nodes shifted by $-0.5\,h$, which will be noted by fractional in m indices. The boundary condition $q(0) = 0 \;\forall\; \eta^j$ is given in the form $q^j_{-1/2} + q^j_{1/2} = 0 \;\forall\; j \leqslant 0$. For the radial part of the Laplace operator at the internal node ρ_m $(0 < m < M)$, we will use the usual approximation

$$\Delta^h f_m = \frac{1}{\rho_m} \frac{1}{h_r} \left(\rho_{m+1/2} \frac{f_{m+1} - f_m}{h_r} - \rho_{m-1/2} \frac{f_m - f_{m-1}}{h_r} \right),$$

and on the axis of symmetry with $\rho_0 = 0$, its limit for even functions:
$$\Delta^h f_0 = \frac{4}{h^2}(f_1 - f_0).$$

8.3.4. Difference method III in the nonlinear case

The considered algorithm is based on relations (8.35)–(8.36) and can be represented as

$$\tilde{R}\mathbf{x}^j = \mathbf{y}^j - \mathbf{G}^n(\hat{\mathbf{x}}^j), \tag{8.37}$$

where $\hat{\mathbf{x}}^j = (q^{j+1}, \varphi^j, \psi^j, \gamma^j)^T$. This record means that only the function q^{j+1} is used to represent the nonlinear terms from the well-known, that is, previously counted layer with respect to the variable η. All other unknowns enter into (8.37) from the time layer j, which makes the 'almost' scheme completely implicit. Of course, this lowers the formal order of approximation in the variable τ to the first and, possibly, imposes a restriction on stability. But all the calculations at step j are carried out using explicit formulas with the exception of a one-time solution of a problem of the form (8.34).

We give an algorithm for implementing the proposed scheme in steps:

1) the calculation of ψ_m^j and φ_m^j:

$$\frac{\psi_m^{j+1} - \psi_m^j}{\tau} - \beta_m^{j+1} = 0, \quad \varphi_m^j = \frac{\psi_m^j - \Delta\psi_m^j}{1 - \psi_m^j};$$

2) calculating γ_m^j :

$$\gamma_m^j = \left(\frac{\left| a_m^j \right|^2}{4} + \frac{1}{2}\left[\left(\psi_m^j \right)^2 + \left(\frac{q_{m-1/2}^{j+1} + q_{m+1/2}^{j+1}}{2} \right)^2 \right] \right) \frac{1}{1 - \psi_m^j};$$

3) calculation of the right-hand side for the problem (8.34).

$$g_m = \varphi_m^j \psi_m^j + \varphi_m^j \gamma_m^j -$$

$$- \frac{1}{\rho_m h} \left(\rho \frac{\partial}{\partial \eta} q\varphi \bigg|_{\rho = \rho_{m+1/2}} - \rho \frac{\partial}{\partial \eta} q\varphi \bigg|_{\rho = \rho_{m-1/2}} \right) \bigg|_{t = t^{j+1}};$$

4) solution of the grid problem (8.34) – definition of the auxiliary function σ_m:

$$(I - \Delta^h)\sigma_m = g_m;$$

5) calculating β_m^j :

$$\frac{\beta_m^{j+1} - \beta_m^j}{\tau} + \gamma_m^j + \psi_m^j + \sigma_m = 0;$$

6) calculation on the shifted grid $q_{m-1/2}^j$:

$$q_{m-1/2}^j + \frac{\beta_m^j - \beta_{m-1}^j}{h} + q_{m-1/2}^{j+1} \frac{\varphi_m^j + \varphi_{m-1}^j}{2} = 0.$$

All calculations for the time layer with the number j are completed. If it is necessary to increase the formal order of approximation in the variable η from the first to the second, repeat the steps 2–6.

Let us emphasize the feature of the difference method III: the computational complexity of the algorithm is sharply reduced while maintaining the formal order of accuracy inherent in method II. Newton's method in combination with the matrix tridiagonal algorithm here gave way to the non-iterative scalar tridiagonal algorithm, which in numerical experiments led to a reduction in real computational costs by about an order, that is, 10 times.

8.4. Projection method

The section for solving a system of nonlinear partial differential equations describing a plasma three-dimensional axially symmetric wake wave deals with the projection (spectral) method. Its constructive basis differs significantly from the above finite difference method schemes.

8.4.1. Setting the problem in a convenient form

Recall the basic formulation of the problem. Equations are conveniently written as

$$\frac{\partial \psi}{\partial \eta} = \beta,$$

$$\frac{\partial \beta}{\partial \eta} + \frac{1}{\rho} \frac{\partial}{\partial \rho} \rho \frac{\partial q}{\partial \eta} = \Delta \gamma - \varphi \gamma + 1,$$

$$0 = q\varphi + \frac{\partial \beta}{\partial \rho}, \tag{8.38}$$

$$0 = \varphi \psi - \Delta \psi + 1,$$

$$0 = \gamma \psi + \frac{1}{2}\left[\psi^2 + q^2 + 1 + \frac{|a|^2}{2} \right].$$

The system of equations (8.38) is presented in a dimensionless form with respect to the unknown functions: γ, β, $\psi = q_z - \gamma$, $\varphi = N/\gamma$, $q = p_r/mc$; for the radial part of the Laplace operator, the notation $\Delta = \frac{1}{\rho} \frac{\partial}{\partial \rho}\left(\rho \frac{\partial}{\partial \rho} \right)$ is used.

Additional conditions are required to uniquely identify the desired functions in the domain

$$\Omega = \left\{ (\rho, \eta) : 0 \leqslant \rho \leqslant R, Z_e \leqslant \eta \leqslant Z_s \right\}$$

We describe their structure, having preliminarily formulated the requirements for the concordance of the parameters of the pulse and the size of the domain Ω.

The physical formulation of the problem provides that the pulse cross section is substantially smaller than the transverse size of the domain occupied by the plasma, that is, $\exp(-R^2 / \rho_*^2) \ll 1$. The longitudinal characteristics of the region Ω were chosen from other considerations: $\exp(-Z_s^2 / l_*^2) \ll 1$, and the value of Z_e was determined only by research objectives and available computational resources. Recall that the variable η in the problem changes in the direction of decreasing, i.e., integration over it is carried out in the opposite direction.

Thus, the specified agreement of the parameters actually means the modeling of the following situation. Before the pulse ($\eta \geq Z_s$) there is an unperturbed plasma, that is, when $0 \leq \rho \leq R$ the initial conditions are given:

$$\psi(\rho, Z_s) = -1, \quad \beta(\rho, Z_s) = 0, \quad q(\rho, Z_s) = 0,$$
$$\varphi(\rho, Z_s) = 1, \quad \gamma(\rho, Z_s) = 1.$$

Then (with $Z_e \leq \eta \leq Z_s$) it changes its structure in accordance with the solution of system (8.38), however, in the cross section the pulse is localized so that the boundary values of the functions (with $\rho = R$) can be considered unperturbed, i.e. corresponding state of rest:

$$\psi(R, \eta) = -1, \quad \beta(R, \eta) = 0, \quad \varphi(R, \eta) = 1, \quad \gamma(R, \eta) = 1;$$

on the axis (at $\rho = 0$), due to the axial symmetry of the problem, the following conditions are posed:

$$\frac{\partial \psi}{\partial \rho}(0, \eta) = \frac{\partial \beta}{\partial \rho}(0, \eta) = \frac{\partial \gamma}{\partial \rho}(0, \eta) = \frac{\partial \varphi}{\partial \rho}(0, \eta) = q(0, \eta) = 0.$$

It should be noted that in such a formulation, the initialization of weakly and strongly nonlinear wake waves in a plasma now depends only on the values of the pulse parameter a_*.

8.4.2. Description of the projection method

To construct an approximate solution in the form of single series with coefficients depending only on the variable η, it is necessary to define two sets of basis functions $\{Y_k(\rho)\}_{k=1}^{K}$ and $\{Z_k(\rho)\}_{k=1}^{K}$, which is associated with different evenness of the desired functions with respect to the axis of symmetry, $\rho = 0$. Let λ_k, $k = 1, 2, \ldots$ be the zeros of the Bessel function $J_0(t)$, that is, $J_0(\lambda_k) = 0$. Then the functions $J_0(\lambda_k t)$, $k = 1, 2, \ldots$ are orthogonal on the segment $[0, 1]$ with weight t (see, for example, [52]):

$$\int_0^1 t J_0(\lambda_k t) J_0(\lambda_l t)\, dt = 0 \quad \text{at} \quad k \neq l.$$

We make the change of variables $t = \rho/R$ and take into account the factor for orthonormality on the segment $[0, R]$, as a result we get the first set of basis functions:

$$Y_k(\rho) = \frac{\sqrt{2}}{R\,|J_1(\lambda_k)|} J_0(\lambda_k \frac{\rho}{R}), \quad k = 1,2,\ldots.$$

The absolute value in the denominator is chosen for reasons of positiveness of the normalization coefficient. Note that all functions $Y_k(\rho)$ are even with respect to the axis $\rho = 0$, that is, they satisfy the relation $\dfrac{\partial Y_k(\rho)}{\partial \rho}\Big|_{\rho=0} = 0$. In addition, they vanish when $\rho = R$. This makes possible representations that take into account the initial and boundary conditions:

$$\psi(\rho,\eta) = -1 + \sum_{k=1}^{K} \psi_k(\eta) Y_k(\rho), \quad \beta(\rho,\eta) = \sum_{k=1}^{K} \beta_k(\eta) Y_k(\rho),$$

$$\gamma(\rho,\eta) = 1 + \sum_{k=1}^{K} \gamma_k(\eta) Y_k(\rho), \quad \varphi(\rho,\eta) = 1 + \sum_{k=1}^{K} \varphi_k(\eta) Y_k(\rho).$$

For convenience, the background (unperturbed) values of variables are clearly identified as separate terms.

To construct a similar decomposition of the function $q(\rho, \eta)$, odd basis functions are required. We use the well-known [52] relations: $J_0'(t) = -J_1(t)$ and $J_{-n}(t) = (-1)^n J_n(t)$. This leads to the orthogonality property –

$$\int_0^1 t J_1(\lambda_k t) J_1(\lambda_l t)\, dt = 0 \quad \text{at} \quad k \neq l,$$

and, respectively, to the second set of basis functions orthonormal on $[0, R]$

$$Z_k(\rho) = -\frac{\sqrt{2}}{R\,|J_1(\lambda_k)|} J_1(\lambda_k \frac{\rho}{R}), \quad k = 1,2,\ldots.$$

As a result, we have a representation

$$q(\rho,\eta) = \sum_{k=1}^{K} q_k(\eta) Z_k(\rho).$$

Now we can obtain a system of differential–algebraic equations for $k = 1, 2, ..., K$ with respect to the unknown coefficients depending only on the variable η. We introduce auxiliary notation for integrals:

$$YYY(k,l,s) = \int_0^R \rho Y_k(\rho) Y_l(\rho) Y_s(\rho) d\rho,$$

$$YZZ(k,l,s) = \int_0^R \rho Y_k(\rho) Z_l(\rho) Z_s(\rho) d\rho,$$

and determine the values $\varkappa_k = \lambda_k / R$. Considering that

$$Y_k'(\rho) = \varkappa_k Z_k(\rho), \quad \Delta Y_k(\rho) = -\varkappa_k^2 Y_k(\rho),$$

substitution of the constructed expansions into the differential equations (8.38) and projecting on the basis functions using the orthogonality conditions allows to write the desired system in the form

$$\psi_k' = \beta_k,$$

$$\beta_k' - \varkappa_k q_k' = -[1 + \varkappa_k^2]\gamma_k - \varphi_k - \sum_{l,s=1}^{K} \varphi_l \gamma_s YYY(k,l,s),$$

$$0 = q_k + \varkappa_k \beta_k + \sum_{l,s=1}^{K} q_l \varphi_s YZZ(s,l,k), \tag{8.39}$$

$$0 = [1 + \varkappa_k^2] \psi_k - \varphi_k + \sum_{l,s=1}^{K} \varphi_l \psi_s YYY(k,l,s),$$

$$0 = -\gamma_k + \sum_{l,s=1}^{K} \psi_l \gamma_s YYY(k,l,s) +$$

$$+ \frac{1}{2} \sum_{l,s=1}^{K} \psi_l \psi_s YYY(k,l,s) + \frac{1}{2} \sum_{l,s=1}^{K} q_l q_s YZZ(k,l,s) + a_k.$$

Here the impulse influence is defined through the right side

$$a_k(\eta) = \frac{1}{4}\int_0^R \rho \,|\,a(\rho,\eta)\,|^2\, Y_k(\rho)d\rho.$$

Recall that to close the system, it is necessary to add homogeneous initial conditions for $\eta = Z_s$:

$$\psi_k(Z_s) = \beta_k(Z_s) = q_k(Z_s) = \varphi_k(Z_s) = \gamma_k(Z_s) = 0,$$

for their coordination (in the algebraic part of the system), the choice of the parameter Z_s is responsible, so that $|a_k(Z_s)| \ll 1$.

8.4.3. Numerical implementation of the projection method

To calculate λ_k (the zeros of the Bessel functions $J_0(t)$), a fractional rational approximation was used from [150]. The coefficients of the polynomials given in it are written in the IEEE arithmetic standard and deliver the required values with double precision, which is convenient when using different FORTRAN-77 compilers. The approximate values of the Bessel functions $J_0(t)$ and $J_1(t)$ we determined using machine-independent programs from the SPECFUN package [2], which is a development of the FUNPACK package [128]. The approximations implemented in these programs theoretically provide at least 18 valid significant digits in decimal arithmetic. In reality, test calculations of the values of $J_0(\lambda_k)$ at $1 \le k \le 300$ showed that their absolute deviations from zero do not exceed the value $0.4 \cdot 10^{-14}$.

The approximate calculation of integrals

$$\int_0^R \rho \exp\{-2(\rho/\rho_*)^2\} Y_k(\rho)d\rho,$$

included as factors in the right-hand sides of $a_k(\eta)$ and integrals YYY (k, l, s), YZZ (k, l, s), which are coefficients of the system (8.39), was carried out using the DQAG program from the QUADPACK package [156], part of the library SLATEC [128]. In order to test the program, we checked the orthonormal relations on the segment $[0, 1]$ with the weight function ρ for the basis functions $Y_k(\rho)$ and $Z_k(\rho)$ constructed above. The calculations showed that for $1 \le k \le 300$ the absolute deviations of the tested relations do not exceed the value of $0.6 \cdot 10^{-13}$.

Numerical integration of a system of first-order differential-algebraic equations of the form $Ly' = f(x, y)$ was carried out using the RADAU5 program from the monograph [138], which implements an implicit 5th-order Runge–Kutta method with accuracy control at the step. The main advantage of the program is the ability to solve systems with large values of the differentiation index or perturbation index, which determines the stiffness of the problem [138]. In our case, the index of the linearized (decomposing into independent blocks) system is 3, and in the nonlinear case it can reach $3K$, where K is the number of terms in the decomposition of the desired functions.

8.5. Numerical experiments and comparison methods

The main purpose of the calculations in the section is to study the dependence of the electron density function N on the coordinates ρ and η for various values of the pulse parameters. Such a statement is connected with the fact that the concept of wake wave destruction in a plasma is based on the concept of 'wave breaking', which is associated with the singularity (going to infinity) of the function $N = \varphi\gamma$. For this, the following presentation is convenient. The system of equations of the basic formulation (7.32) can be reduced to a single equation for the function ψ [146], which is the potential of the forces acting on an electron moving along the OZ axis with the speed of light c. Moreover, all functions from (8.38) are expressed in terms of ψ and its derivatives. For example, the desired electron density is

$$N(\rho,\eta) = \frac{1}{2}\left\{ F(\rho,\eta)\left(1 + \psi^{-2}(\rho,\eta)\right) + G^2(\rho,\eta) / F(\rho,\eta) \right\}, \qquad (8.40)$$

where $F(\rho, \eta) = 1 - \Delta\psi$, $G(\rho,\eta) = \dfrac{\partial^2 \psi}{\partial\rho\partial\eta}$ are the axial and radial flows of electrons in the coordinate system moving with the pulse, respectively.

Numerical calculations carried out using both the three finite difference methods and the projection method gave a good quantitative coincidence and showed the following. The wake wave excited by the source retains its structure only at a certain distance behind the pulse. As the phase front of the wave moves away from it, it becomes more and more curved and non-uniformly deformed, which leads to the formation of peaks of electron density off-axis. Further, these emerging peaks grow and surpass in amplitude the maxima of the density peaks on the axis. This growth is highly non-linear and leads to large absolute values tending to infinity. Typical for calculations is the following picture, corresponding to the parameters of the pulse $a_* = 1.2$, $\rho_* = 1.8$, $l_* = 3.5$. Figure 8.1 *a* and *b* show the isolines of the electron density $N(\rho, \eta)$ and perturbations relative to the background value of the potential $\delta\psi(\rho, \eta)$ (so that $\psi = \delta\psi - 1$) with $0 \le \rho \le 1$ and $45 \le \eta \le 65$. Distances in the longitudinal variable are measured from the centre of the pulse.

Figure 8.2 shows the isolines of the axial $F(\rho, \eta)$ and radial $G(\rho, \eta)$ electron fluxes in the same boundaries and scales.

It is easy to see that the main part of the electron density determines the axial flow of electrons, that is, the first term in the formula (8.40). Its shape follows the form of density, and the absolute values are more than an order of magnitude greater than the values of the radial flow. This means that the formation of singular density values is not associated with the vanishing of the quantities ψ and F, but with the growth of the transversa part of the Laplacian from the potential.

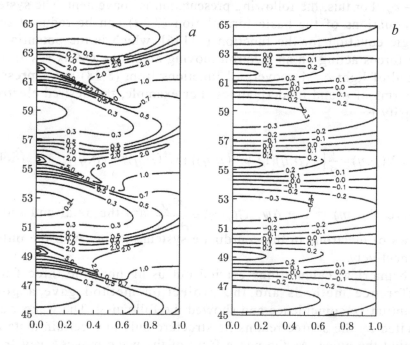

Fig. 8.1. Isolines of electron density $N(\rho, \eta)$ (*a*) and perturbations of the potential $\delta\psi(\rho, \eta)$ (*b*).

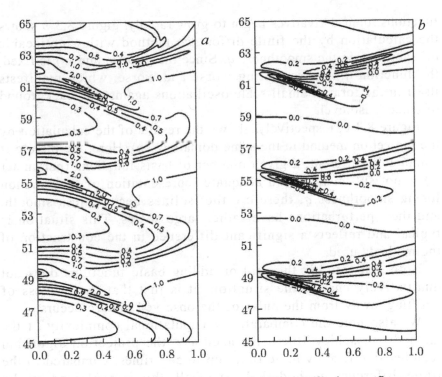

Fig. 8.2. Isolines of axial $F(\rho, \eta)$ (*a*) and radial $G(\rho, \eta)$ (*b*) electron fluxes.

Also of certain interest is the difference between the two approaches (finite-difference and projection) in calculating functions that lose spatial smoothness with distance from the laser pulse. A good illustration of this is Fig. 8.3 at which for $a_* = 1.1$ (the other parameters of the pulse are determined as in the previous calculation) the dynamics of the function q describing the radial velocity of electrons is presented. The parameters of the finite difference method are the grid steps $h_r = 5 \cdot 10^{-3}$, $h_z = 5 \cdot 10^{-4}$; and the parameters of the projection method are the number of basis functions $K = 200$ and the relative accuracy of the numerical integration of the system of differential algebraic equations $\delta_{rel} = 10^{-6}$.

At the initial stage, when $0 \le \eta \le 20$ (see Fig. 8.3 a), $q(\rho, \eta)$ has sufficient smoothness, i.e., spatial derivatives of the required order that are continuous and limited to small quantities. Therefore, the values of the function itself, calculated using both approaches, are practically the same (a difference may be observed in the 4th/5th significant digit).

Further along with the distance from the source, the wake wave is destroyed, which significantly affects the behaviour of all functions: their derivatives begin to grow rapidly. Figure 8.3 *b* shows the calculation by the finite-difference method with a noticeable deterioration in the smoothness q. Since the grid (h_r and h_z) is fixed, the function $q(\rho, \eta)$ is getting worse and worse, which manifests itself in the form of small-scale oscillations against the background of smooth isolevel.

Figure 8.3 *c*, respectively, shows the results of the calculation by the projection method in the same domain (ρ, η). Here, the picture is completely different: a given number of basis functions (parameter K) is not sufficient for an adequate representation of the function losing smoothness q, therefore the isolines become less smooth, and their perturbations have rather large scales. This situation is typical and reflects a significant difference in the construction of the methods used.

Note that refining the grid or adding basic functions does not qualitatively change the situation: it is just that the process of 'shifting' away from the pulse of the observed picture occurs.

We also note the comparative computational complexity of the methods: to achieve the same accuracy, the projection algorithm requires significantly more time (up to 5–7 times) compared to the finite difference method II. First of all, this is explained by the automatic choice of the integration step with respect to the variable η in the domains of large gradients of the unknown functions. The best of the algorithms for modelling wake waves based on the hydrodynamic model considered in the chapter is the difference method III (linearization method). The worst is the projection (spectral) method. In terms of computational efficiency, method II is about an order of magnitude (10 times). In turn, method I is about 5 times less than method II. It is about the same set of parameters of the differential algebraic problem, as well as grid parameters.

Method III (linearization method) has the following differences from the other three algorithms:

1) the possibility to carry out calculations until breaking. Other algorithms stop working, as a rule, after forming the first off-axis maximum of the electron density in the solution. Further progress in time, the same as for the method of linearization, was possible to carry out only for method II, but at very high computational costs.

2) the superiority in efficiency of the best of the other methods (difference method II) is about an order of magnitude, even when the solution of the problem has not yet lost its smoothness;

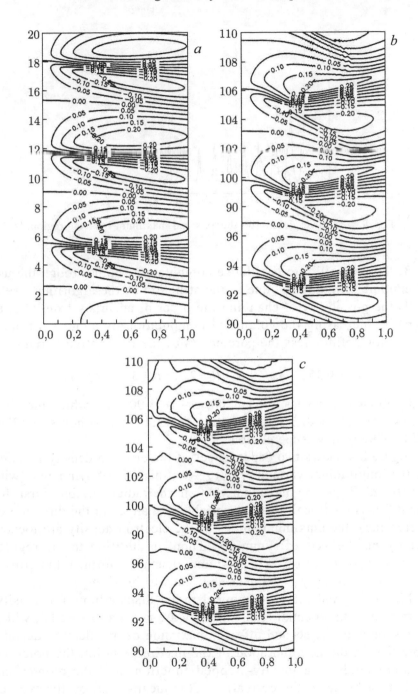

Fig. 8.3. Isolines of the radial velocity of electrons $q(\rho, \eta)$, obtained by difference and projection methods.

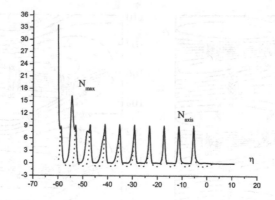

Fig. 8.4. Electron density on the axis and maximum over the region.

3) the possibility to parallelize computations both through the use of explicit formulas and due to the structure of the algorithm itself;

4) the possibility of changing the order of accuracy from $O(\tau)$ to $O(\tau^2)$ by increasing the volume of computation by about 1.6 times.

Fix for definiteness the parameters of the illustrative variant

$$a_* = 0.352, \quad \rho_* = 0.6, \quad R = 2.7, \quad l_* = 3.5, \quad Z_s = 11$$

and grid characteristics $h = 1/3200$, $\tau = 1/64\ 000$, for which the plots below are obtained. The calculation took about two hours at SKIF MSU 'Chebyshev' supercomputer.

Figure 8.4 shows two dependences of the electron density function on the longitudinal variable η: N_{axis} – on the axis of symmetry (with $\rho = 0$) and N_{max} – the maximum value in the radial variable. First, for about 8 periods (recall that the variable η changes in the direction of decreasing), the maximum values of the electron density are located strictly on the axis of symmetry. Then, in addition to the regular axial, a sequence of off-axis maximum density is formed, the growth of which is strongly non-linear and leads to breaking.

Figure 8.5 and 8.6 show typical radial distributions for density, momentum and potential gradient. They are taken at $\eta \approx -54.4$, which corresponds to the second off-axis maximum of the electron density. Note that in the neighborhood of the maximum point, the potential gradient tends to a locally-stepped function, and the momentum tends to a break in the derivative. This means that, by the type of destruction of the solution, the breaking of oscillations corresponds to a 'gradient catastrophe', since the functions themselves (momentum,

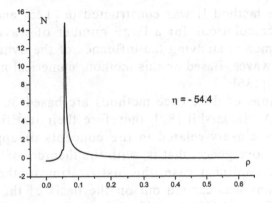

Fig. 8.5. Radial density distribution – second off-axis maximum.

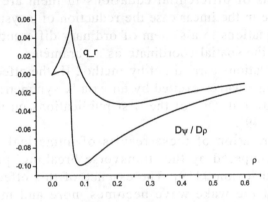

Fig. 8.6. Radial distribution of momentum and potential gradient – second off-axis maximum.

potential gradient) remain bounded, and their derivatives have a singularity.

8.6. Bibliography and comments

The chapter is fully devoted to the construction of various numerical algorithms for modelling plasma wake waves in the basic formulation.

Difference method I was constructed in [109]; based on it, a large number of calculations were carried out for modeling wake waves ([108, 110–112], etc.). The first studies were aimed at studying the smooth part of the solution of the problem, i.e., on the regular behaviour of the wave, suitable for accelerating the electrons injected into it.

Difference method II was constructed in [125] and focused on performing calculations for a large number of wave periods, in particular, aimed at studying the influence of the dynamics of ions in wakefield waves. Based on this method, numerical modelling was performed in [136].

Constructions of difference methods are based on the classical theory of A.A. Samarskii [82], therefore their justification in the linear case is closely related to the concepts of approximation, stability, and covergence, that is, with the theorem of A.F. Filippov [22]. For the nonlinear case, the justification of the stability of both algorithms was carried out on the basis of the principle of 'frozen coefficients' [20, 41, 90]. For this purpose, the difference approximations of differential equations in them are designed to fully reproduce in the linear case the reduction of a system of partial differential equations to a system of ordinary differential equations depending on the spatial coordinate as a parameter.

In the calculations carried out by method II, the effect of off-axis breaking of wake waves excited by an axially symmetric laser pulse was first discovered. One of the first publications on this topic are the works [11, 49].

The interpretation of these results of numerical experiments was greatly hampered by the 'transverse breaking' point of view prevailing at that time [119]. The essence of this effect is that the phase front of the wake wave becomes more and more concave [118] due to a decrease in the frequency of plasma oscillations with an increase in their amplitude [160]. As a result, the wavelength of the oscillations turns out to be maximum on the axis, where the amplitude of the wake wave is maximum, and decreases with distance from the axis. If we assume that the motion of an electron liquid in the process of oscillations occurs orthogonally to the phase front [119], then with increasing distance from the pulse, the radial displacement of the elements of the fluid volume increases. In the end, this should lead to the fact that all the elements of the fluid that started at a certain distance from the axis converge on the axis and a phenomenon of the type of intersection of particle trajectories, which is called the 'transverse breaking', will occur. In fact, this effect was considered only from qualitative considerations and is not justified in any way by the properties of the solution of a system of differential hydrodynamic equations. As a result, it was based on numerical experiments carried out by the particle-in-cell method. It should be noted that the computing resources available to the authors of the idea of 'transverse breaking' at that time were very limited. As a result, coarse grid parameters did not allow the identification of off-axis breaking, which led to the conclusion that the wake wave would break on the axis of symmetry [32–34, 119], and with a feature such as a 'dovetail' [14].

An attempt at more accurate consideration on the basis of hydrodynamic equations was made in [144], where, however, such assumptions were made that did not allow the authors to derive a certain criterion for the wake wave breaking.

Numerous consultations with major experts in the field of numerical methods and solving applied problems, primarily with Academician N.S. Bakhvalov and Professor V.I. Lebedev, led to the idea that a fundamentally different numerical method should be constructed to solve the same problem. It is possible that an approximation to a solution obtained on a different ideological basis will not be related to the shortcomings (specificity) of the finite difference method, but will be fully adequate in relation to the original problem. As a result of this, a projection (spectral) method was constructed in [126]. Its construction is closely related to the idea of Galerkin approximations [71] in systems of special functions [5]. The spectral method turned out to be very laborious: at computational costs roughly comparable to the difference method I. However, calculations based on it confirmed that a new (off-axis) effect of breaking axially symmetric wake waves was detected. To substantiate it, an analytical theory of weakly nonlinear wake waves excited by a sharply focused laser pulse [93] was constructed, and a special study was carried out for their axial solutions [100]. The results obtained for these auxiliary problems are fully consistent with the numerical experiments for the basic formulation.

It should be clarified that, in principle, axial wake wave breaking is possible. For example, in chapter 5 it is shown that when taking into account the dynamics of ions, plasma oscillations can break both on the axis of symmetry of the problem and outside it. The specific situation is determined by the parameters of the pulse exciting the oscillations. Field experiments also confirm the possibility of axial breaking. However, the essence of the problem here is that in order to explain this phenomenon, it is necessary to involve models that take into account some additional factors, and not just the dynamics of electrons. And if the conclusions are based solely on the results of numerical experiments, it should be borne in mind that the effect of off-axis breaking appears at a distance of the order of several percent of the pulse width ρ_* from the axis of symmetry. In other words, to capture an effect in the calculation, thousands of points along each spatial variable are required. This means that even for spatially two-dimensional modeling it is necessary to use modern supercomputers.

In addition to the results of this chapter, for a detailed study of the effect of off-axis breaking, numerical, analytical, and asymptotic studies of simpler problems associated with free plasma oscillations were carried out, which are described in the first part of the monograph. Based on the results obtained in [67, 68], difference method III (linearization method) was constructed, which currently seems to be the most suitable for simulating wake waves in the hydrodynamic approximation.

As noted above, the reason for the study of the breaking of plasma oscillations was not only qualitative but also quantitative coincidence of the electron density with a similar characteristic during the propagation of the wake wave. This coincidence was first noted in [45] for the case of nonrelativistic oscillations; later it was established when relativism was taken into account [123]. We describe the result of the comparison, following [45].

Consider a typical simulation of cylindrical oscillations from Chapter 4 with parameters

$$a_* = 0.365, \quad \rho_* = 0.6, \quad d = 2.7, \quad h = 1/1600, \quad \tau = 1/32000,$$

in which the functions of velocity and electric field are unknown, and the electron density is calculated by the formula (4.3). The images in Figs. 8.7 and 8.8 of the electron density function $N(\rho, \theta)$ in the subdomain $0 \le \rho \le 0.1$, $0 \le \theta \le 35$ are represented.

The figures clearly show the whole process of changing the shape of oscillations. The analogy with the mountain relief, when the peak is located on the ridge, seems to be very successful here. Located on the axis $\rho = 0$, the peak-like density perturbations along with the 'ridge' are first simply rotated counterclockwise (as viewed in the direction of their propagation), and then the bending of the 'ridges' in the vicinity of regular maxima is added to the rotation. This stage corresponds to Fig. 8.7. Finally, the electron density has an off-axis maximum comparable in magnitude with the axial ones. Here, the angle of rotation and the magnitude of the bend are so large that it is difficult to reflect the described effect on the general plan of disturbances (a higher level of resolution is required). Therefore, the

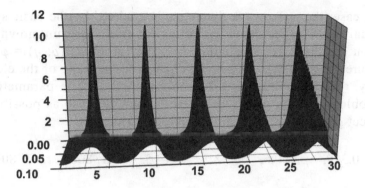

Fig. 8.7. Regular development of oscillations.

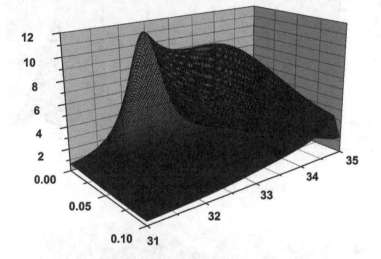

Fig. 8.8. First off-axis maximum.

'ridge' with two maxima is shown separately in Fig. 8.8, and for clarity, the angle of observation is somewhat changed.

This moment is fundamental: further from period to period, the smoothness of the function $N(\rho, \theta)$ deteriorates significantly and after about one or two periods the electron density goes to infinity due to the intersection of the trajectories of neighboring electrons. In practice, stopping of calculations in the Eulerian variables occurs due to the appearance of negative density values near very large density values (on the order of several thousand).

The considered process has a pronounced analogue in the behavior of wake waves excited in a plasma by a short high-power laser pulse.

In this case, the wake wave dynamics is modeled by the basic system of equations (7.32), in which other functions are unknown. The electron density here is calculated by the formula $N(\rho, \eta) = \varphi \cdot \gamma$.

Figures 8.9 and 8.10 shows the change in the shape of the electron density, calculated by the difference method II. The parameters of the problem and the grid steps were chosen as close as possible for a correct comparison with the above experiment:

$$a_* = 0.360, \quad \rho_* = 0.6, \quad R = 2.7, \quad l_* = 3.5, \quad h = 1/800, \quad \tau = 1/8000.$$

The electron density varies in the same qualitative scenario, but

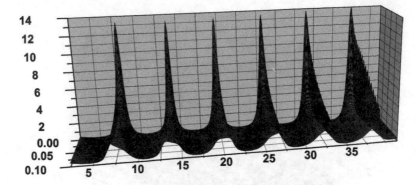

Fig. 8.9. Regular wave dynamics.

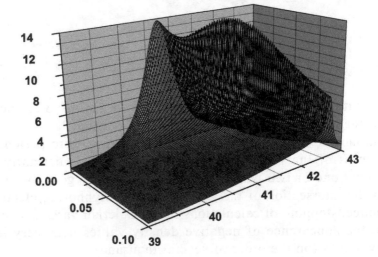

Fig. 8.10. Off-axis maximum on the wave.

slightly more smoothly. In Fig. 8.9 it can be seen that axial peaks retain their shape longer and, accordingly, the 'ridges' rotate more slowly. The off-axis maximum of the electron density in the wake wave is observed one period farther from the centre of the pulse as compared with the similar picture arising in the process of oscillations, and its absolute value itself is somewhat larger than the nearest regular maximum. However, here too, the appearance of an off-axis maximum indicates that the wake wave will soon break down. Its external signs are completely similar to those considered above: from period to period, there is a sharp increase in the off-axis maximum, near which negative values of the electron density appear. Refining the grid parameters very briefly, as in the case of electron oscillations, prolongs the lifetime of the solution being sought.

Such coincidence of decisions is not accidental. The qualitative and quantitative similarity in the behaviour of the main characteristic (electron density) in the study of the plasma oscillations and wake waves means the similarity of the physical phenomena themselves, which ultimately helped to understand the causes and mechanisms of destruction of both processes. Although it should be noted that modeling three-dimensional axially symmetric wake waves in a plasma is a significantly more complex problem than simulating actually one-dimensional plasma oscillatios.

Additional research

The chapter discusses various problems of modelling plasma wake waves, whose settings differ from the basic ones. Numerical algorithms for solving these problems, as well as the very properties of these solutions are of independent interest.

9.1. Axial wake wave solution

In the section, the formulation of the problem is carried out for studying the axial solution during the propagation of a wake wave excited by a short laser pulse. On the axis of symmetry, a comparison is made between the solution of the complete (in the basic formulation) of the problem and the axial solution. The obtained results are in favour of off-axis wake wake.

9.1.1. Formulation of the 'truncated' problem

In contrast to the basic formulation of the problem (system of equations (7.31), (7.32) together with a set of initial and boundary conditions (7.36)–(7.38)), which we will call 'complete', we formulate a 'truncated' statement describing the axial solution, i.e., solution in the neighborhood of the symmetry axis of the basic problem.

In a dimensionless form, the fundamental relations of the basic formulation are written with respect to the unknown functions γ, β, ψ, φ, q:

$$a(\rho,\eta) = a_* \exp\left\{-\frac{\rho^2}{\rho_*^2} - \frac{\eta^2}{l_*^2}\right\},$$

$$\frac{\partial \psi}{\partial \eta} = \beta, \quad \frac{\partial \beta}{\partial \eta} + \frac{1}{\rho}\frac{\partial}{\partial \rho}\rho\frac{\partial q}{\partial \eta} = \Delta \gamma - \varphi\gamma + 1, \tag{9.1}$$

$$q\varphi + \frac{\partial \beta}{\partial \rho} = 0, \quad \varphi\psi - \Delta\psi + 1 = 0, \quad 2\gamma\psi + \psi^2 + q^2 + 1 + \frac{|a|^2}{2} = 0,$$

where for the radial part of the Laplace operator, the notation
$\Delta = \dfrac{1}{\rho}\dfrac{\partial}{\partial\rho}\left(\rho\dfrac{\partial}{\partial\rho}\right)$ is used.

The basis for the derivation of the 'truncated' equations are the initial conditions describing the unperturbed plasma, as well as the boundary conditions on the axis of symmetry of the problem. We assume that before the pulse ($\eta \leq Z_s$), i.e., when $0 \leq \rho \leq R$,

$$\psi(\rho,Z_s) = -1, \beta(\rho,Z_s) = 0, q(\rho,Z_s) = 0, \varphi(\rho,Z_s) = 1, \gamma(\rho,Z_s) = 1, \tag{9.2}$$

and also on the axis (for $\rho = 0$), due to the axial symmetry of the problem, it is defined:

$$\frac{\partial\psi}{\partial\rho}(0,\eta) = \frac{\partial\beta}{\partial\rho}(0,\eta) = \frac{\partial\gamma}{\partial\rho}(0,\eta) = \frac{\partial\varphi}{\partial\rho}(0,\eta) = q(0,\eta) = 0. \tag{9.3}$$

Taking into account the type of boundary conditions (9.3), that is, in fact, the conditions of evenness or oddness of the unknown functions with respect to the axis of symmetry $\rho = 0$, we will consider the axial solution, that is, the solution of the following structure:

$$\psi(\rho,\eta) = \psi_0(\eta) + \psi_2(\eta)\rho^2, \quad \beta(\rho,\eta) = \beta_0(\eta) + \beta_2(\eta)\rho^2,$$
$$\varphi(\rho,\eta) = \varphi_0(\eta) + \varphi_2(\eta)\rho^2, \quad \gamma(\rho,\eta) = \gamma_0(\eta) + \gamma_2(\eta)\rho^2, \tag{9.4}$$
$$q(\rho,\eta) = q_1(\eta)\rho.$$

Here, the terms of higher order in the variable ρ are omitted, that is, only the lower terms of the expansion are retained, and the subscript of the function indicates that it is a multiplier depending on the variable η, with the corresponding degree of the variable ρ.

Substituting the form of solutions (9.4) into the system (9.1) and equating the coefficients with the same powers ρ, we obtain the differential algebraic system

$$\psi_0' = \beta_0, \quad \psi_2' = \beta_2, \quad \beta_0' + 2q_1' = 4\gamma_2 - \varphi_0\gamma_0 + 1,$$
$$\varphi_0\psi_0 - 4\psi_2 + 1 = 0, \quad 2\gamma_0\psi_0 + \psi_0^2 + 1 + S^2(\eta)/2 = 0, \quad (9.5)$$
$$q_1\varphi_0 + 2\beta_2 = 0, \quad \gamma_2\psi_0 + \gamma_0\psi_2 + \psi_0\psi_2 + q_1^2/2 - S^2(\eta)/(2\rho_*^2) = 0,$$

where the function $S(\eta)$ (analogue of the function $a(\rho, \eta)$ from the complete problem) has the form

$$S(\eta) = a_* \exp\left\{-\frac{\eta^2}{l_*^2}\right\}.$$

Note that here, unlike (9.1), the number of equations (seven) does not coincide with the number of unknowns (eight), therefore, to close the system, you need to add some additional relation (differential or algebraic). For this purpose, we use the approximation for small a_*. Note that the small values of a_* mean only the smallness of the values of the functions ψ_0, β_0, γ_0 on the axis, but not the smallness of the radial derivatives, that is, ψ_2, β_2, γ_2, q_1. Moreover, the axial density oscillations are determined primarily by the values of ψ_2 and can exceed the background value by an order of magnitude, since they substantially depend not only on a_*, but also on ρ_*.

Assuming that a_* is small, we represent the solution of the complete system (9.1) in the form

$$\psi(\rho,\eta) = -1 + \tilde{\psi}(\rho,\eta) + o(\tilde{\psi}), \quad \beta(\rho,\eta) = 0 + \tilde{\beta}(\rho,\eta) + o(\tilde{\beta}),$$
$$\varphi(\rho,\eta) = 1 + \tilde{\varphi}(\rho,\eta) + o(\tilde{\varphi}), \quad \gamma(\rho,\eta) = 1 + \tilde{\gamma}(\rho,\eta) + o(\tilde{\gamma}), \quad (9.6)$$
$$q(\rho,\eta) = 0 + \tilde{q}(\rho,\eta) + o(\tilde{q}).$$

Here, for convenience, the background (unperturbed) values of functions are explicitly identified. In this case, up to terms of the second order of smallness, from the first four equations (9.1) we will have:

$$\frac{\partial \tilde{\psi}}{\partial \eta} = \tilde{\beta}, \quad \frac{\partial \tilde{\beta}}{\partial \eta} + \frac{1}{\rho} \frac{\partial}{\partial \rho} \rho \frac{\partial \tilde{q}}{\partial \eta} = \Delta \tilde{\gamma} - \tilde{\varphi} - \tilde{\gamma},$$

$$\tilde{q} + \frac{\partial \tilde{\beta}}{\partial \rho} = 0, \quad \tilde{\varphi} - \tilde{\psi} - \Delta \tilde{\psi} = 0.$$

After the simple transformations obtained from the resulting system (see section 7.3.2, and also [109]), the equation

$$\frac{\partial \tilde{\beta}}{\partial \eta} + \tilde{\psi} + \tilde{\gamma} = 0.$$

It is formally valid for arbitrary values of ρ and η, but here we will use it only on the axis for the closure of system (9.5). Applying the representation equation (9.4) and (9.6) to the obtained equation, we obtain the missing equation for $\rho = 0$:

$$\beta_0' + \psi_0 + \gamma_0 = 0. \tag{9.7}$$

Considering the structure of the obtained system (9.5), (9.7), it is convenient to separate the problem of determining axial solutions into two subproblems. The first is the subproblem for the independent determination of the functions themselves on the axis:

$$\psi_0' = \beta_0, \quad \beta_0' + \psi_0 + \gamma_0 = 0, \quad 2\gamma_0 \psi_0 + \psi_0^2 + 1 + S^2(\eta)/2 = 0, \tag{9.8}$$

endowed with initial conditions at $\eta = Z_s$:

$$\psi_0(Z_s) = -1, \quad \beta_0(Z_s) = 0, \tag{9.9}$$

and the second is a subproblem for radial derivatives with already defined functions $\psi_0, \beta_0, \gamma_0$:

$$\psi_2' = \beta_2, \quad q_1 \varphi_0 + 2\beta_2 = 0, \quad \varphi_0 \psi_0 - 4\psi_2 + 1 = 0,$$
$$2q_1' = 4\gamma_2 - \varphi_0 \gamma_0 + 1 + \varphi_0 + \gamma_0, \tag{9.10}$$
$$\gamma_2 \psi_0 + \gamma_0 \psi_2 + \psi_0 \psi_2 + q_1^2/2 - S^2(\eta)/(2\rho_*^2) = 0,$$

with their initial conditions

$$\psi_2(Z_s) = 0, \quad q_1(Z_s) = 0, \tag{9.11}$$

Thus, the problem of determining axial solutions in this case reduces to integrating equations (9.8), (9.10) on the time interval $Z_e \leq \eta \leq Z_s$, starting from the initial conditions (9.9), (9.11). The desired value of the electron density function on the axis is expressed by the formula

$$N(\rho = 0, \eta) = \varphi_0 \gamma_0 = -\frac{\gamma_0}{\psi_0}(1 - 4\psi_2). \tag{9.12}$$

9.1.2. Numerical algorithm for solving the 'truncated' problem

We introduce a uniform grid

$$\eta_k = k\tau, \quad K_{min} \leq k \leq K_{max}, \quad (K_{max} - K_{min})\tau = Z_s - Z_e$$

and write discrete analogs of the formulated subproblems using the notation $f^k = f(\eta_k)$ for grid functions. For problem (9.8), (9.9) we will have calculation formulas

$$\frac{\psi_0^{k+1} - \psi_0^k}{\tau} = \beta_0^{k+1},$$

$$\gamma_0^k = -\frac{(\psi_0^k)^2 + 1 + S^2(\eta_k)/2}{2\psi_0^k}, \tag{9.13}$$

$$\frac{\beta_0^{k+1} - \beta_0^k}{\tau} + \psi_0^k + \gamma_0^k = 0$$

and the initial conditions

$$\psi_0^{K_{max}} = -1, \quad \beta_0^{K_{max}} = 0. \tag{9.14}$$

Note that the scheme (9.13), (9.14) is explicit, since the integration over the variable η is in the direction of its decrease.

For the numerical solution of equations (9.10), it is convenient to first exclude from the system the functions β_2, γ_2, φ_0 and bring it to the form

$$\psi_2' = \frac{q_1}{2\psi_0}(1 - 4\psi_2),$$

$$q_1' + \frac{q_1^2}{\psi_0} = \frac{1}{2}\left[1 + \psi_0 + \gamma_0 + \frac{\gamma_0}{\psi_0}(1 - 4\psi_2)\right] - \frac{2}{\psi_0}\left[\psi_2(\gamma_0 + \psi_0) - \frac{S^2(\eta)}{2\rho_*^2}\right],$$

and only then write the appropriate calculation scheme:

$$\frac{\psi_2^{k+1} - \psi_2^k}{\tau} = \frac{q_1^{k+1}}{2\psi_0^{k+1}}(1 - 4\psi_2^{k+1}),$$

$$\frac{q_1^{k+1} - q_1^k}{\tau} + \frac{q_1^k}{\psi_0^k} = \frac{1}{2}\left[1 + \psi_0^k + \gamma_0^k + \frac{\gamma_0^k}{\psi_0^k}(1 - 4\psi_2^k)\right] -$$

$$-\frac{2}{\psi_0^k}\left[\psi_2^k(\gamma_0^k + \psi_0^k) - \frac{S^{\perp}(\eta_k)}{2\rho_*^2}\right],$$

$$(9.15)$$

providing it with necessary initial conditions

$$\psi_2^{K_{max}} = 0, \quad q_1^{K_{max}} = 0. \tag{9.16}$$

In contrast to (9.13), (9.14), the scheme (9.15), (9.16) is implemented explicitly-implicitly: the values of ψ_2^k from the first relation are determined explicitly, and the values of q_1^k from the second (non-linear) are calculated using the classical Newton method. In this case, the iterations stop when the residual modulus does not exceed δ_{newt}. In the calculations, the initial approximation was taken from the neighboring time step, and the value δ_{newt} was assumed to be approximately equal to τ^2.

It should be noted that the analysis of the numerical algorithm, carried out for small variations (perturbations) of the solution, suggests that the linearized scheme has a second order of accuracy with respect to the parameter τ and is equivalent to the standard 'leapfrog' scheme.

9.1.3. Calculation results

We first consider the results of calculations of the 'complete' problem with parameters

$$a_* = 0.088, \quad \rho_* = 0.15, \quad l_* = 3.5, \quad Z_s = 11.0.$$

Figures 9.1 and 9.2 show plots of the functions $\beta_f = \tilde{\beta}(0, \eta)$ and $\psi_f = \tilde{\psi}(0, \eta)$, from which it follows that the assumption of their smallness on the axis is quite suitable, since the deviations from the background values of these functions in modulus do not exceed the values of $2 \cdot 10^{-3}$ and $3 \cdot 10^{-3}$ respectively. In addition, the values of the function $\tilde{\gamma}(\rho, \eta)$, even in the vicinity of the centre of the pulse,

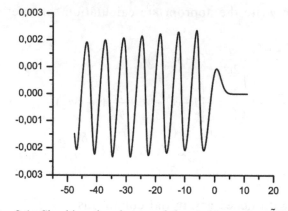

Fig. 9.1. Checking the closure of the problem $\beta_f = \tilde{\beta}(0,\eta)$.

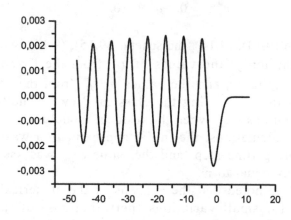

Fig. 9.2. Checking the closure of the problem $\psi_f = \tilde{\psi}(0,\eta)$.

do not exceed in magnitude the values of $2 \cdot 10^{-3}$. All this testifies to the correctness of using relation (9.7) for the closure of system (9.5).

We now turn to the analysis of Table 9.1. The column $N_{f,1}$ for the indicated values of the parameter η, gives the extremal values of the electron density function at $\rho = 0$, taken from the calculation of the 'complete' problem by the difference method II with the following grid parameters: $h_1 = 0.5 \cdot 10^{-3}$, $\tau_1 = 0.5 \cdot 10^{-4}$.

We note that in this case, when $\eta = -46.6$, the grid wake wave has already ceased to exist (breaking took place). The next column $N_{f,2}$ contains the values of the same function, but for the grid parameters $h_2 = h_1/2$, $\tau_2 = \tau_1/4$. It is easy to see that in the smoothness domain of the solution (that is, with $\eta \geq -43.7$), the deviations of the extremal values do not exceed 0.5%, in other words, the convergence in the

Table 9.1

η	$N_{f,1}$	$N_{f,2}$	$N_{a,2}$
−02.0	0.2231	0.2231	0.2231
−05.4	11.483	11.536	11.584
−08.3	0.2556	0.2556	0.2554
−11.3	11.500	11.547	11.602
−14.2	0.2556	0.2555	0.2553
−17.2	11.500	11.539	11.588
−20.1	0.2555	0.2555	0.2553
−23.1	11.505	11.536	11.566
−26.0	0.2555	0.2555	0.2555
−29.0	11.510	11.531	11.535
−31.9	0.2556	0.2556	0.2558
−34.9	11.505	11.517	11.495
−37.8	0.2558	0.2558	0.2561
−40.7	11.514	11.527	11.541
−43.7	0.2560	0.2561	0.2565
−46.6	Breaking	11.542	11.554
−49.6		Breaking	0.2568
−52.5			11.569
−55.5			0.2570
−58.4			11.585
−61.4			0.2572
−64.3			11.601

grid has already been achieved with satisfactory accuracy. In this case, the breaking of the grid wake wave occurs somewhat farther from the centre of the pulse (when $\eta \geq -47.6$), which, in turn, indicates that the grid parameters are sufficiently small, but not about their artificially low. And, finally, the column $N_{a,2}$ contains the electron density values obtained from solving the 'truncated' problem using the grid parameter τ_2. Here, deviations from the values of $N_{f,2}$ also do not exceed 0.5%, which in explicit form indicates the correctness of the proposed method for simulating the axial solutions.

We give an estimate of the reduction in the amount of calculations. In this case, the compared values are not very easy to calculate. The fact is that at each step in the variable η, the Newton method is used to solve both problems ('complete' and 'truncated').

Moreover, for the 'complete' problem, it is realized by the matrix 3×3 Thomas algorithm [83], and for the 'truncated' problem – by scalar formulas (for one equation). Therefore, even if we neglect the slowdown of convergence for a system of approximately $M = 2600$ equations (as compared to convergence for one equation), the lower estimate for reducing the volume of computational work looks like $M \times K_p > 20\ 000$ times (i.e. $K_p \geq 8$). Here, K_p denotes the constant in the asymptotics of the tridiagonal matrix algorithm, depending on whether the coefficients are variable or constant [83]. Note that such a 'small' contraction is very noticeable in practical calculations, due to the significant computational complexity of calculations of wake waves. For example, even the computationally simple version of the 'complete' task requires about 10 hours of processor time when using an Intel Core2 Duo Wolfdale CPU (2.66 GHz). Therefore, the definition of only an axial solution for 2–3 s on the same processor is of particular interest in multiple calculations for solving auxiliary optimization problems.

At the end of the section we present in Figs. 9.3–9.8 illustrations of axial solutions, considering their content. It should be noted that the radial derivatives, i.e., q_1, ψ_2, are at least three orders of magnitude greater than the values of the axial perturbations of the functions β_0, ψ_0, γ_0. Therefore, for small a_*, only the derivatives determine the electron density values on the axis $\rho = 0$. This emphasizes the validity of the closing relation (9.7) for extracting axial solutions from system (9.5).

Of course, in the behaviour of the functions $\tilde{\beta}(0,\eta)$, $\tilde{\psi}(0,\eta)$ from the solution of the 'complete' problem (Figs. 9.1 and 9.2) and the functions $\beta_0(\eta)$, $\tilde{\psi_0} = \psi_0(\eta)+1$ from the solution of the 'truncated' problems (Figs. 9.5 and 9.6) has some differences. In particular, the axial solutions are more pronounced periodic in comparison with solutions of the 'complete' problem on the axis. However, due to the smallness of their absolute values, it is very difficult to argue that the deviation from the periodic behaviour is not due to the approximation errors in the radial variable.

We note that axial solutions exist at longer distances from the centre of the pulse compared to solutions of the 'complete' problem; This is a strong argument in favour of off-axis breaking of wake waves. The fact is that earlier there were publications containing an alternative point of view, namely, about axial breaking (see, for example, [32] and works cited there), due to the occurrence of a 'dovetail' type [14] in the 'comlete' problem solution. Based on the

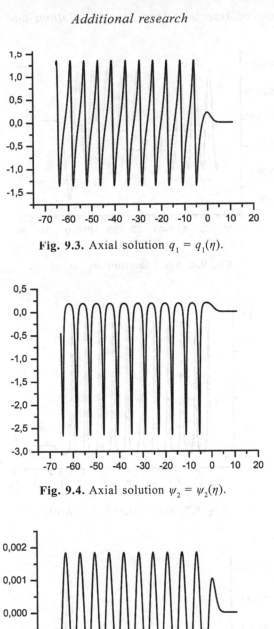

Fig. 9.3. Axial solution $q_1 = q_1(\eta)$.

Fig. 9.4. Axial solution $\psi_2 = \psi_2(\eta)$.

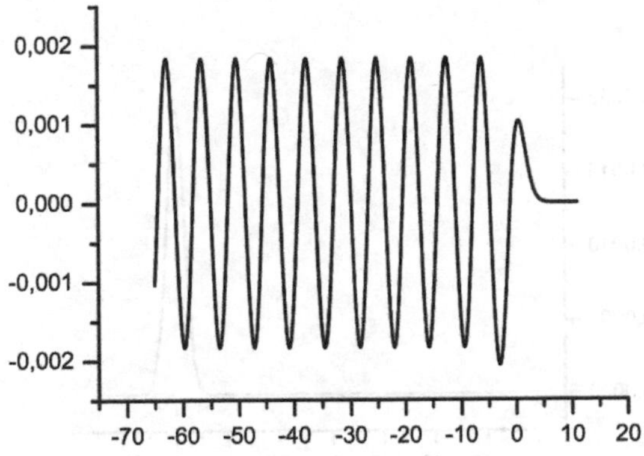

Fig. 9.5. Axial solution $\beta_0 = \beta_0(\eta)$.

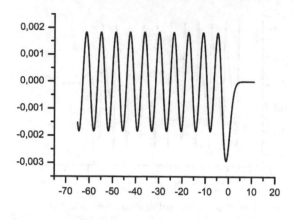

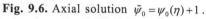

Fig. 9.6. Axial solution $\tilde{\psi}_0 = \psi_0(\eta) + 1$.

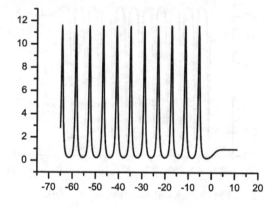

Fig. 9.7. Axial solution $N = N(\eta)$.

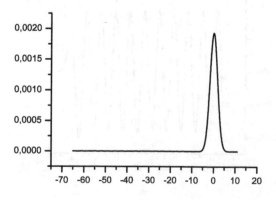

Fig. 9.8. Axial solution $\gamma_0 = \gamma_0(\eta) - 1$.

numerical calculations presented in this section, we can conclude that the long-lived (at least several periods) wake waves breakj outside the axis of symmetry of the problem.

Recall that the calculations of the 'complete' problem given in this section were carried out using the difference method II.

9.2. Accounting for the dynamics of ions in the wake wave

To solve a system of nonlinear partial differential equations describing the three-dimensional dynamics of ions and electrons in a plasma wake wave, this section discusses a difference scheme and an iterative algorithm for its implementation. The results of numerical simulation showed that a domain with a low ion density (the so-called plasma channel) is formed behind the short laser pulse, leading to the destruction of the wake wave.

9.2.1. Problem statement in physical variables

As a rule, considering the propagation of laser pulses and the excitation of wake waves, we use a number of simplifying assumptions. For example, they neglect the motion of ions whose mass is more than three orders of magnitude greater than the mass of electrons. However, in experiments (see, for example, [135, 147, 148, 171]) performed at high intensities of laser radiation, fluxes of fast ions were detected. In addition, a very popular article [35] is very useful on this topic. Obviously, the description of such effects requires taking into account the motion of ions, therefore the basic formulation of the problem from section 7.3 requires upgrading by adding the necessary equations, the initial and boundary conditions for ions.

Let us complicate model (7.1)–(7.8) at the expense of the equation describing the dynamics of the complex amplitude of the laser field a (the so-called envelope) [133]:

$$\left(\frac{2i\omega}{c^2}\frac{\partial}{\partial t}+2ik\frac{\partial}{\partial z}+\Delta_\perp\right)a=\frac{\omega_{p0}^2}{c^2}\left(\frac{n_e}{n_{e_0}\gamma}-1\right)a, \tag{9.17}$$

where ω, k is the frequency and the wave number of the laser pulse propagating along the axis OZ; n_{e_0} is the electron concentration in front of the laser pulse, where the plasma is considered neutral $(e_i n_{i_0} + e n_{e_0} = 0)$, $\omega_{p0} = \sqrt{4\pi n_{e_0} e^2 / m_e}$ is the plasma frequency,

$\Delta_\perp = \dfrac{\partial^2}{\partial x^2} + \dfrac{\partial^2}{\partial y^2}$ is the transverse part of the Laplace operator.

In the axially symmetric case, all quantities depend only on z, r, and t. Introducing new independent variables

$$\eta = \frac{\omega_{p0}}{c}(z - ct), \quad \rho = \frac{\omega_{p0}}{c}r, \quad \theta = \omega_{p0}t$$

and new dimensionless functions

$$q = \frac{\mathbf{p}_e}{m_e c}, \quad \mathbf{w}^e = \frac{\mathbf{v}_e}{c}, \quad \mathbf{w}^i = \frac{\mathbf{v}_i}{c}, \quad \epsilon = \frac{e\mathbf{E}}{m_e c \omega_{p0}}, \quad v_e = \frac{n_e}{n_{e_0}}, \quad v_i = \frac{n_i}{n_{i_0}},$$

from equations (7.1)–(7.8), (9.17) in the quasi-static approximation, when

$$\frac{\partial}{\partial\theta} \ll \frac{\partial}{\partial\eta}, \quad \varepsilon = \frac{\omega_{p0}}{\omega} \ll 1, \quad |\mathbf{w}_i| \ll 1,$$

as was shown earlier in section 7.2.2, it is possible to obtain the initial set of equations for numerical simulation.

As a result, the mathematical description of the problem under study is as follows. In the unbounded domain Ω:

$$\Omega = \left\{ (\rho, \eta, \theta) : 0 \le \rho < \infty, -\infty < \eta < \infty, 0 \le \theta \right\}$$

considered a system of equations

$$\frac{\partial}{\partial\eta}\left[v_e\left(w_z^e - 1\right) \right] + \frac{1}{\rho}\frac{\partial}{\partial\rho}\left(\rho v_e w_r^e \right) = 0,$$

$$\frac{\partial q_z}{\partial\eta} = -\varepsilon_z + \frac{\partial\gamma}{\partial\eta},$$

$$\frac{\partial q_r}{\partial\eta} = -\varepsilon_r + \frac{\partial\gamma}{\partial\rho},$$

$$\frac{\partial}{\partial\eta}\left[v_i\left(w_z^i - 1\right) \right] + \frac{1}{\rho}\frac{\partial}{\partial\rho}\left(\rho v_i w_r^i \right) = 0,$$

$$\frac{\partial w_z^i}{\partial \eta} = \varepsilon_z \delta, \quad \frac{\partial w_r^i}{\partial \eta} = \varepsilon_r \delta,$$

$$\gamma = \sqrt{1 + q_r^2 + q_z^2 + \frac{|a|^2}{2}},$$

$$-\frac{\partial \varepsilon_z}{\partial \eta} = -v_e w_z^e + v_i w_z^i - \frac{1}{\rho}\frac{\partial}{\partial \rho}\left[\rho\left(\frac{\partial q_r}{\partial \eta} - \frac{\partial q_z}{\partial \rho}\right)\right],$$

$$-\frac{\partial \varepsilon_r}{\partial \eta} = -v_e w_r^e + v_i w_r^i + \frac{\partial}{\partial \eta}\left(\frac{\partial q_r}{\partial \eta} - \frac{\partial q_z}{\partial \rho}\right),$$

$$\left\{2\mathbf{i}\frac{1}{\varepsilon}\frac{\partial}{\partial \theta} + \frac{1}{\rho}\frac{\partial}{\partial \rho}\left(\rho\frac{\partial}{\partial \rho}\right)\right\}a = \left(\frac{v_e}{\gamma} - 1\right)a$$

under additional conditions:

$$v_{e,i}(\rho, \eta \to \infty) = 1, \quad \frac{\partial v_{e,i}}{\partial \rho}(\rho = 0) = \frac{\partial v_{e,i}}{\partial \eta}(\eta \to \infty) = 0,$$

$$w_z^{e,i}(\rho \to \infty) = w_z^{e,i}(\eta \to \infty) = q_z(\rho \to \infty) = q_z(\eta \to \infty) = 0,$$

$$w_r^{e,i}(\rho \to \infty) = w_r^{e,i}(\eta \to \infty) = q_r(\rho \to \infty) = q_r(\eta \to \infty) = 0,$$

$$w_r^{e,i}(\rho = 0) = q_r(\rho = 0) = \frac{\partial w_r^{e,i}}{\partial \rho}(\rho = 0) = \frac{\partial q_r}{\partial \rho}(\rho = 0) = 0,$$

$$a(\rho \to \infty) = \frac{\partial a}{\partial \rho}(\rho = 0) = 0,$$

$$\varepsilon_r(\rho \to \infty) = \varepsilon_r(\eta \to \infty) = \varepsilon_r(\rho = 0) = 0,$$

$$\varepsilon_z(\rho \to \infty) = \varepsilon_z(\eta \to \infty) = 0,$$

$$a(\theta = 0) = a_* \exp\left\{-\frac{\rho^2}{\rho_*^2} - \frac{(\eta - \eta_*)^2}{l_*^2}\right\}.$$

Here a, $v_{e,i}$, γ, $w_r^{e,i}$, $w_z^{e,i}$, q_r, q_z, ε_r, ε_z are respectively the dimensionless complex amplitude of laser radiation, the density of electrons and ions, the Lorentz factor, the projections of the velocity vectors of the electrons and ions on the ρ and η axes, projections on the same axes of the momentum vectors of electrons and the electric field are the desired functions $\delta = -\dfrac{e_i m_e}{e m_i} \ll 1, \varepsilon, a_*, \rho_*, \eta_*, l_*$ are the dimensionless parameters of the problem.

9.2.2. Statement of the problem in convenient variables

The above relations after a number of auxiliary transformations, such as those described in detail in section 7.2.4, take the following form:

in the domain $\Omega = \left\{ (\rho, \eta, \theta) : 0 \leq \rho \leq R, Z_e \leq \eta \leq Z_s, 0 \leq \theta \leq T_{\max} \right\}$

find a solution to the system of equations

$$\frac{2\mathbf{i}}{\varepsilon \theta_*} \frac{\partial a}{\partial \theta} + \Delta a + (1 - \varphi)a = 0, \tag{9.18}$$

$$\frac{\partial \psi}{\partial \eta} - \beta = 0, \tag{9.19}$$

$$\varphi \psi - \Delta \psi + v(1 - v) = 0, \tag{9.20}$$

$$\frac{\partial \beta}{\partial \eta} + \frac{1}{\rho} \frac{\partial}{\partial \rho} \rho \frac{\partial q}{\partial \eta} - \Delta \gamma + \varphi \gamma - v = 0, \tag{9.21}$$

$$q\varphi + \frac{\partial \beta}{\partial \rho} - vw = 0, \tag{9.22}$$

$$2\gamma \psi + \psi^2 + q^2 + 1 + \frac{|a|^2}{2} = 0, \tag{9.23}$$

$$v + \delta(1 + \psi) = 0, \tag{9.24}$$

$$\frac{\partial w}{\partial \eta} - \delta \left(\frac{\partial \gamma}{\partial \rho} - \frac{\partial q}{\partial \eta} \right) = 0, \tag{9.25}$$

$$\frac{\partial}{\partial\eta}[v(v-1)] + \frac{1}{\rho}\frac{\partial}{\partial\rho}(\rho vw) = 0, \qquad (9.26)$$

for boundary and initial conditions:
  at $0 \le \rho \le R$

$$\psi(\rho, \eta = Z_s) = -1, \quad \beta(\rho, \eta = Z_s) = 0,$$
$$v(\rho, \eta = Z_s) = 1, \quad w(\rho, \eta = Z_s) = 0, \quad v(\rho, \eta = Z_s) = 0;$$

with $Z_e \le \eta \le Z_s$:

$$\psi(\rho = R, \eta) = -1, \quad q(\rho = R, \eta) = 0,$$
$$\varphi(\rho = R, \eta) = 1, \quad a(\rho = R, \eta) = 0,$$
$$v(\rho = R, \eta) = 1, \quad w(\rho = R, \eta) = 0, \quad v(\rho = R, \eta) = 0,$$
$$\frac{\partial\psi}{\partial\rho}(\rho = 0, \eta) = \frac{\partial\gamma}{\partial\rho}(\rho = 0, \eta) = \frac{\partial a}{\partial\rho}(\rho = 0, \eta) = 0,$$
$$\frac{\partial v}{\partial\rho}(\rho = 0, \eta) = \frac{\partial v}{\partial\rho}(\rho = 0, \eta) = w(\rho = 0, \eta) = q(\rho = 0, \eta) = 0;$$

when $\theta = 0$:

$$a(\rho, \eta) = a_* \exp\left\{-\frac{\rho^2}{\rho_*^2} - \frac{(\eta - \eta_*)^2}{l_*^2}\right\} \qquad (9.27)$$

Here we use the notation

$$\Delta = \frac{1}{\rho}\frac{\partial}{\partial\rho}\left(\rho\frac{\partial}{\partial\rho}\right), \psi = q_z - \gamma, \beta = \frac{\partial\psi}{\partial\eta},$$

$$\varphi = \frac{v_e}{\gamma}, v_i = v, q_r = q, w_z^i = v, w_r^i = w.$$

We note that the possibility of transforming a solution domain from unbounded in ρ and η variables to a bounded one is connected with the following circumstances. The localization of the solution in ρ follows from the form of the initial pulse at $\theta = 0$ and the boundedness of the time interval of interest. With respect to the variable η, the zero boundary conditions are satisfied with a sufficient accuracy at a finite distance from the centre of the pulse. Since the Cauchy problem requires conditions on one side (boundary) of the domain, the position of the other boundary is determined only by the interests of the study.

Specifying specific quantities R, Z_e, Z_s, T_{max}, ε, δ, a_*, ρ_*, η_*, l_*, θ_* completes the statement of the problem. We give for them the typical values used in the calculations:

$$R = 2 \div 5, \ Z_s = -15, \ Z_e = 250, \ T_{max} = 1500, \ \varepsilon = 1/90,$$
$$a_* = 0.5 \div 2.0, \ \rho_* = 1.5 \div 5.0, \ \eta_* = 0, l_* = 2 \div 5, \ \delta = 0 \div 0.01, \ \varepsilon\theta_* = 2/9 \div 8/9.$$

We note some features of the formulated problem, which are important for further discussion. Despite the fact that all the required functions depend on the three variables ρ, η, θ, the dependence on θ in equations (9.19)–(9.26) is implicit: in the form of the last term $|a|^2/2$ on the left side of equation (9.23) – the functions of (ρ, η, θ), in the same way, in equation (9.18) (in the form of a coefficient, the function $\varphi(\rho, \eta, \theta)$) there is a dependence on η. For this reason, the relevant parts of the problem solving algorithm are described further as for the case of only two independent variables. It should also be noted that the variable η changes in the direction of decreasing, i.e., integration over it is conducted in the opposite direction.

9.2.3. Solution method

In the domain Ω, we construct a uniform grid over all variables with steps h_ρ, h_z and τ so that

$$\rho_m = mh_r, \ 0 \le m \le M; \quad \eta_j = jh_z, \ 0 \le j \le N; \quad \theta_n = n\tau, \ 0 \le n,$$

and proceed to the description of the difference scheme for the above problem. In this case, we will use for grid functions notation of the form $f(\rho_m, \eta_j, \theta_n) = f^n_{m,j}$, omitting some indices where this should not cause confusion; the functions q and w will be defined in the nodes shifted by $-0.5\, h_\rho$, which is marked by fractional indices with respect to m.

We present the difference equations for determining the grid functions inside the domain for $j = N - 1$, $N - 2$, ..., 0; $m = 1, 2, ..., M - 1$:

$$\frac{2\mathbf{i}}{\varepsilon\theta_*}\frac{a_{m,j}^{n+1}-a_{m,j}^{n}}{\tau}+\Delta a_{m,j}^{n+1}+\left(1-\varphi_{m,j}\right)a_{m,j}^{n+1}=0, \tag{9.28}$$

$$\frac{\psi_{m,j+1}-\psi_{m,j}}{h_z}-\beta_{m,j+1}=0, \tag{9.29}$$

$$v_{m,j}=\delta(1\mid\psi_{m,j}), \tag{9.30}$$

$$\frac{v_{m,j+1}(v_{m,j+1}-1)-v_{m,j}(v_{m,j}-1)}{h_z}+$$
$$+\frac{1}{\rho_m h_r}\left(\rho_{m+\frac{1}{2}}w_{m+\frac{1}{2},j+1}\frac{v_{m,j+1}+v_{m+1,j+1}}{2}-\right.$$
$$\left.-\rho_{m-\frac{1}{2}}w_{m-\frac{1}{2},j+1}\frac{v_{m,j+1}+v_{m-1,j+1}}{2}\right)=0, \tag{9.31}$$

$$\varphi_{m,j}\psi_{m,j}-\Delta\psi_{m,j}+v_{m,j}(1-v_{m,j})=0, \tag{9.32}$$

$$\frac{\beta_{m,j+1}-\beta_{m,j}}{h_z}-\Delta\gamma_{m,j}+\varphi_{m,j}\gamma_{m,j}-v_{m,j}+$$
$$+\frac{1}{\rho_m h_r}\left(\rho_{m+\frac{1}{2}}\frac{q_{m+\frac{1}{2},j+1}-q_{m+\frac{1}{2},j}}{h_z}-\rho_{m-\frac{1}{2}}\frac{q_{m-\frac{1}{2},j+1}-q_{m-\frac{1}{2},j}}{h_z}\right)=0, \tag{9.33}$$

$$q_{m-\frac{1}{2},j}\frac{\varphi_{m,j}+\varphi_{m-1,j}}{2}+\frac{\beta_{m,j}-\beta_{m-1,j}}{h_r}=w_{m-\frac{1}{2},j}\frac{v_{m,j}+v_{m-1,j}}{2}, \tag{9.34}$$

$$2\gamma_{m,j}\psi_{m,j}+\psi_{m,j}^2+\left(\frac{q_{m-\frac{1}{2},j}+q_{m+\frac{1}{2},j}}{2}\right)^2+1+\frac{\mid a_{m,j}^{n+1}\mid^2}{2}=0, \tag{9.35}$$

$$\frac{w_{m-\frac{1}{2},j+1}-w_{m-\frac{1}{2},j}}{h_z}+\delta\frac{q_{m-\frac{1}{2},j+1}-q_{m-\frac{1}{2},j}}{h_z}=\delta\frac{\gamma_{m,j}-\gamma_{m-1,j}}{h_r}. \tag{9.36}$$

Above for the radial part of the Laplace operator

$$\Delta=\frac{1}{\rho}\frac{\partial}{\partial\rho}\left(\rho\frac{\partial}{\partial\rho}\right)$$

in the node ρ_m, $m \geq 1$, the usual approximation is used:

$$\Delta f_m = \frac{1}{\rho_m}\frac{1}{h_r}\left(\rho_{m+1/2}\frac{f_{m+1}-f_m}{h_r} - \rho_{m-1/2}\frac{f_m-f_{m-1}}{h_r}\right).$$

Next, we write out the difference analogs of the initial and boundary conditions – the necessary relations for determining the grid functions on the boundary Ω:

when $j = N$; $m = 0, 1, \ldots, M$:

$$\psi_{m,N} = -1, \quad \beta_{m,N} = q_{m-\frac{1}{2},N} = 0, \quad \varphi_{m,N} = 1, \quad \gamma_{m,N} = 1,$$

$$v_{m,N} = 1, \quad w_{m-\frac{1}{2},N} = 0, \quad \nu_{m,N} = 0,$$

when $m = M$; $j = N, N - 1, \ldots, 0$:

$$\psi_{M,j} = -1, \quad \varphi_{M,j} = 1, \quad \beta_{M,j} = q_{M-\frac{1}{2},j} = \gamma_{M,j} = a_{M,j}^{n+1} = 0,$$

$$v_{M,j} = 1, \quad w_{M-\frac{1}{2},j} = \nu_{M,j} = 0,$$

when $m = 0$; $j = N, N - 1, \ldots, 0$:

$$\frac{2\mathbf{i}}{\varepsilon\theta_*}\frac{a_{0,j}^{n+1}-a_{0,j}^n}{\tau} + \Delta a_{0,j}^{n+1} + \left(1-\varphi_{0,j}\right)a_{0,j}^{n+1} = 0,$$

$$\frac{\psi_{0,j+1}-\psi_{0,j}}{h_z} - \beta_{0,j+1} = 0,$$

$$\varphi_{0,j}\psi_{0,j} - \Delta\psi_{0,j} + v_{0,j}(1-v_{0,j}) = 0,$$

$$\frac{\beta_{0,j+1}-\beta_{0,j}}{h_z} - \Delta\gamma_{0,j} + \varphi_{0,j}\gamma_{0,j} - v_{0,j} + \frac{2}{h_r}\left(\frac{q_{+\frac{1}{2},j+1}-q_{+\frac{1}{2},j}}{h_z} - \frac{q_{-\frac{1}{2},j+1}-q_{-\frac{1}{2},j}}{h_z}\right) = 0,$$

$$q_{-\frac{1}{2},j}\frac{\varphi_{0,j}+\varphi_{1,j}}{2} + \frac{\beta_{0,j}-\beta_{1,j}}{h_r} = w_{-\frac{1}{2},j}\frac{v_{0,j}+v_{1,j}}{2},$$

$$2\gamma_{0,j}\psi_{0,j} + \psi_{0,j}^2 + \left(\frac{q_{-\frac{1}{2},j}+q_{+\frac{1}{2},j}}{2}\right)^2 + 1 + \frac{|a_{0,j}^{n+1}|^2}{2} = 0,$$

$$\frac{v_{0,j+1}(v_{0,j+1}-1)-v_{0,j}(v_{0,j}-1)}{h_z}+\left(v_{0,j+1}+v_{1,j+1}\right)\frac{w_{\frac{1}{2},j+1}-w_{-\frac{1}{2},j+1}}{h_r}=0,$$

$$\frac{w_{-\frac{1}{2},j+1}-w_{-\frac{1}{2},j}}{h_z}+\delta\frac{q_{-\frac{1}{2},j+1}-q_{-\frac{1}{2},j}}{h_z}=\delta\frac{\gamma_{0,j}-\gamma_{1,j}}{h_r},$$

here

$$\Delta f_0=\frac{4}{h_r^2}(f_1-f_0)$$

– the difference analogue of the Laplace operator on the axis of symmetry $\rho = 0$, taking into account the condition of the evenness of the function $f(\rho)$;
  when $n = 0$; $j = N$, $N - 1$, ..., 0; $m = 0, 1, ..., M$:

$$a_{m,j}^0=a_*\exp\left\{-\frac{\rho_m^2}{\rho_*^2}-\frac{(\eta_j-\eta_*)^2}{l_*^2}\right\}.$$

The difference scheme used for the linearized problem has the second order of approximation with respect to the step in ρ and the first in θ, η; stability and, therefore, convergence with the same orders can be obtained for it in the standard way, as in sections 8.1 and 8.2.

Let us consider the implementation of the constructed scheme. Let at time θ_n be known the values of $a_{m,j}^n$ for all m, j from the domain of definition. To solve equation (9.28), an iterative method of the type of compressing mappings was used (s is the iteration number):

$$\frac{2i}{\varepsilon\theta_*}\frac{a_{m,j}^{n+1,s+1}-a_{m,j}^n}{\tau}+\Delta a_{m,j}^{n+1,s+1}+\left(1-\varphi_{m,j}^s\right)a_{m,j}^{n+1,s+1}=0, \tag{9.37}$$

where the initial approximation was chosen as $a_{m,j}^{n+1,0}=a_{m,j}^n$, and $\varphi_{m,j}^s$ is the solution of system (9.29)–(9.36), in which its function was used at the previous iteration for the function $a_{m,j}^{n+1}$, i.e. $a_{m,j}^{n+1,s}$.

The algorithm for solving the system (9.29)–(9.36) was as follows. For each $j = N - 1$, $N - 2$, ..., 0, the functions $\psi_{m,j}$ from equation (9.29), $v_{m,j}$ from (9.30), $v_{m,j}$ from (9.31) and $\varphi_{m,j}$ were determined using explicit formulas. from (9.32); then the three-point nonlinear vector system (9.33)–(9.36) was solved by the Newton method implemented by the matrix (4 × 4) Thomas algorithm. The accuracy was fixed by the maximum in m and all the equations for the residual not exceeding $\delta_{\text{newt}} = 10^{-3}$. As a rule, this corresponded to no more than three iterations.

After applying the described algorithm, we obtained the values of $\varphi_{m,j}^s$ for all j and successive tridiagonal matrix methods with respect to ρ solved equation (9.37) to determine $a_{m,j}^{n+1,s+1}$.

This technique (an inserted combination of explicit formulas and a Newtonian loop) was repeated until

$$\max_{m,j} | a_{m,j}^{n+1,s+1} - a_{m,j}^{n+1,s} | > \delta_{it} = 10^{-3}.$$

After one or two iterations of this type, the sought value of $a_{m,j}^{n+1}$ for the time moment θ_{n+1} was obtained, and the outer loop ended.

Thus, the total time step was a double inserted iterative loop: the inner one was Newtonian for the four functions of the system (9.33)–(9.36), and the outer one was of the type of compressing mappings for the function a using equation (9.37).

It should be noted the usefulness of scaling the desired variables, similar to that described in section 5.2. In the calculations, both variants of the proposed difference scheme were used. Scaling allows to reduce the influence of errors arising due to the finiteness of the iterative processes, especially in multi-period calculations, which make it possible to accurately reflect the influence of the dynamics of heavy ions.

9.2.4. Calculation results

Numerical simulation was carried out for a stationary pulse of the form (9.27) and showed that a short laser pulse ($l_* = 3.5$) effectively excites a wake wave, which, acting on ions, gradually forms behind the laser pulse a domain with a low plasma density (plasma channel) . As one moves away from the pulse, the depth of the channel first increases and then decreases. Starting from a certain moment, the concentration of ions on the axis of symmetry begins to monotonously increase, which eventually leads to the destruction of the wake wave.

Figure 9.9 shows the isolines of the electron (lower half of the figure) and ion (upper half of the figure) densities behind the narrow ($\rho_* = 2$) laser pulse ($a_* = 1$, $\delta = 1/2000$) in the plane (ρ, η). For convenience, the plots in this section show the variable η in the opposite direction.

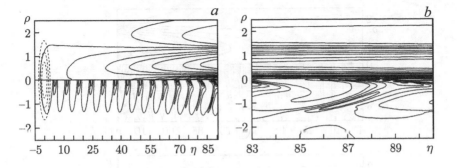

Fig. 9.9. Isolines of electron and ion density functions behind a narrow laser pulse: regular development and wave destruction.

Figure 9.9 *a* shows the wake wave evolution in a wide range of variation of the variable η ($-5 < \eta < 90$). Figure 9.9 *b* shows, on the same scale, in both variables, the domain of destruction of the solution. In this case, the channel has a tubular shape with a maximum radial gradient of ion density near the axis, where small-scale changes in the electron density occur. Here, the wake wave breaks in the axial domain due to a significant increase in the concentration of ions on the axis of symmetry. Similar to the experiments described in section 5.3, one can choose the parameters of the pulse so that the breaking occurs strictly on the axis of symmetry of the problem (see illustrative Figs. 5.6–5.8).

Figure 9.10 shows the same functions as Fig.9.9, but for a wider laser pulse ($\rho_* = 3$, $a_* = 1.2$, $\delta = 1/2000$). In this case, the maximum of the radial gradient of the ion density is removed from the axis, where wake wave breaking occurs. In this case, off-axis breaking of the wake wave takes place, i.e. a situation when the influence of the dynamics of ions is small. A detailed description of the analogue of such breaking over for plane electron–ion oscillations is also given in section 5.3; corresponding illustrations are shown in Figs. 5.3–5.5.

For comparison Fig. 9.11 shows the evolution of a wake wave with the same pulse parameters as in Fig. 9.10, but with infinitely heavy ions ($\delta = 0$). The destruction of the solution and the appearance of small-scale electron density variations occur at much larger distances from the pulse and are associated with the curvature of the wake wave phase front due to a relativistic change in the electron mass and the geometric properties of the formulation of the problem. An analogue of such a breaking in the case of relativistic cylindrical oscillations is described in detail in Chapter 4 (see also paper [46]).

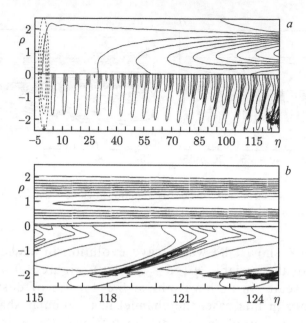

Fig. 9.10. Isolines of electron and ion density functions behind a wide laser pulse: regular development and wave destruction

A more in-depth study of the dynamics of ions, conducted by the method described above, is described in a special work [136].

9.3. Elliptical pulse

In the section for solving a system of nonlinear partial differential equations describing the three-dimensional dynamics of ions and electrons in a plasma wake wave excited by a powerful short laser pulse, the difference scheme and an iterative algorithm for its implementation are considered. A significant difference from the previous considerations is the rejection of the axial symmetry of the problem, which leads to complication of the equations and an increase in the dimension. The results of numerical simulating illustrate the difference in wake waves generated by pulses of circular and elliptical cross-sections.

9.3.1. Formulation of the problem

As a rule, considering the propagation of laser pulses and the excitation of wake waves, a number of simplifying assumptions are made. For example, they neglect the motion of ions whose mass

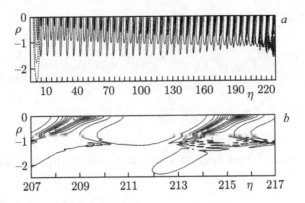

Fig. 9.11. Isolines of the electron density function behind the wide laser pulse with stationary ions: regular development and wave destruction.

is more than three orders of magnitude greater than the mass of the electrons. Another traditional simplification is the assumption of the axial symmetry of the problem. In fact, the ions are slowly but moving, and the distribution of the intensity of radiation in a laser pulse is to some extent different from axially symmetric, which is reflected in the structure of wake waves. In the section, the formulation of the problem is considered, which takes into account both the dynamics of ions and the rejection of the axial symmetry of the problem. This leads to an increase in both the number of equations and the number of independent variables.

Let us complicate the model by the equation describing the dynamics of the complex amplitude of the laser field a – the envelope (see section 9.2.1, and also [133]):

$$\left(\frac{2i\omega}{c^2}\frac{\partial}{\partial t} + 2ik\frac{\partial}{\partial z} + \Delta_\perp \right) a = \frac{\omega_{p0}^2}{c^2}\left(\frac{n_e}{n_{e_0}\gamma} - 1 \right) a, \qquad (9.38)$$

where ω, k is the frequency and the wave number of the laser pulse propagating along the axis OZ; n_{e_0} is the electron concentration before the laser pulse, where the plasma is considered neutral $(e_i n_{i_0} + e n_{e_0} = 0)$ $\omega_{p0} = \sqrt{4\pi n_{e_0} e^2 / m_e}$ is the plasma frequency, $\Delta_\perp = \dfrac{\partial^2}{\partial x^2} + \dfrac{\partial^2}{\partial y^2}$ is the transverse part of the Laplace operator.

Now, introducing new independent variables

$$\eta = \frac{\omega_{p0}}{c}(z - ct), \quad x' = \frac{\omega_{p0}}{c}x, \quad y' = \frac{\omega_{p0}}{c}y, \quad \theta = \omega_{p0}t$$

and new dimensionless functions

$$\mathbf{q} = \frac{\mathbf{p}_e}{m_e c} = (q_x, q_y, q_z)^T, \ \mathbf{v} = \frac{\mathbf{v}_i}{c} = (v_x, v_y, v_z)^T, \ b_z = \frac{eB_z}{m_e c \omega_{p0}},$$

$$v = \frac{n_i}{n_{i_0}}, \ \psi = q_z - \gamma, \ \beta = \frac{\partial \psi}{\partial \eta}, \ \varphi = \frac{v_e}{\gamma},$$

from equations (7.1)–(7.8) in the quasistatic approximation (see section 7.2.2)

$$\frac{\partial}{\partial \theta} \ll \frac{\partial}{\partial \eta}, \quad \varepsilon = \frac{\omega_{p0}}{\omega} \ll 1, \quad |\mathbf{v}| \ll 1$$

the initial for the numerical simulating system of equations follows. As a result, the mathematical description of the problem under study is as follows. In the domain $\Omega \times [0, T_{max}]$, where

$$\Omega = \left\{ (x', y', \eta) : |x'| \leq X_{max}, |y'| \leq Y_{max}, Z_e \leq \eta \leq Z_s \right\}, \quad 0 \leq \theta \leq T_{max},$$

find a solution to the system of equations:

$$\frac{\partial \psi}{\partial \eta} - \beta = 0, \tag{9.39}$$

$$v_z + \delta(1 + \psi) = 0, \tag{9.40}$$

$$\frac{\partial}{\partial \eta} \left[v(v_z - 1) \right] + \frac{\partial}{\partial x'} (v v_x) + \frac{\partial}{\partial y'} (v v_y) = 0, \tag{9.41}$$

$$\varphi \psi - \Delta_\perp \psi + v(1 - v_z) = 0, \tag{9.42}$$

$$\frac{\partial \beta}{\partial \eta} + \frac{\partial}{\partial \eta} \left(\frac{\partial q_x}{\partial x'} + \frac{\partial q_y}{\partial y'} \right) - \Delta_\perp \gamma + \varphi \gamma - v = 0, \tag{9.43}$$

$$q_x \varphi + \frac{\partial \beta}{\partial x'} - v v_x - \frac{\partial b_z}{\partial y'} = 0, \tag{9.44}$$

$$q_y \varphi + \frac{\partial \beta}{\partial y'} - vv_y + \frac{\partial b_z}{\partial x'} = 0, \tag{9.45}$$

$$2\gamma\psi + \psi^2 + q_x^2 + q_y^2 + 1 + \frac{|a|^2}{2} = 0, \tag{9.46}$$

$$\frac{\partial v_x}{\partial \eta} - \delta\left(\frac{\partial \gamma}{\partial x'} - \frac{\partial q_x}{\partial \eta}\right) = 0, \tag{9.47}$$

$$\frac{\partial v_y}{\partial \eta} - \delta\left(\frac{\partial \gamma}{\partial y'} - \frac{\partial q_y}{\partial \eta}\right) = 0, \tag{9.48}$$

$$b_z = \frac{\partial q_x}{\partial y'} - \frac{\partial q_y}{\partial x'}, \tag{9.49}$$

$$\frac{2\mathbf{i}}{\varepsilon}\frac{\partial a}{\partial \theta} + \Delta_\perp a + (1 - \varphi)a = 0. \tag{9.50}$$

The system depends on two numerical coefficients: $\delta = -\dfrac{e_i m_e}{e m_i}$, $\varepsilon = \dfrac{\omega_\rho 0}{\omega}$ and its solution consists of the following 12 functions:

$$\psi, \beta, \varphi, \gamma, v, v_x, v_y, v_z, q_x, q_y, b_z, a,$$

which, due to the specifics of the posing, depend on independent variables in various ways. For example, the first 11 of them are real functions that implicitly depend on the time variable θ and explicitly on the variables x', y', η: they are connected by a closed system (9.39)–(9.49), which includes the value $|a|^2/2$ (see equation (9.46)), depending on time. This means that these unknown functions depend on time as an external parameter. On the other hand, the complex-valued function a, which satisfies equation (9.50), explicitly depends on the variables θ, x', y' and implicitly on η: the dependence on the latter is manifested through the function φ, which enters the equation in the form of a given coefficient.

To uniquely determine the desired functions, additional (initial and boundary) conditions are required. We describe their structure to complete the formulation of the problem. At the initial time $\theta = 0$, the envelope is set

$$a(x', y', \eta) = a_* \exp\left\{ -\frac{(x')^2 + \alpha(y')^2}{\rho_*^2} - \frac{\eta^2}{l_*^2} \right\}, \qquad (9.51)$$

where a_*, α, ρ_*, l_* are dimensionless parameters defining the amplitude and geometric shape of the laser pulse. Note that the value $\alpha = 1$ corresponds to the pulse of a circular cross-section, for the formulation of the problem with which an axially symmetric description is admissible. The centre of the pulse, for simplicity, is at the point (0, 0, 0).

We formulate the requirements for the concordance of the pulse parameters and the size of the domain Ω. The physical formulation of the problem assumes that the pulse cross section is substantially smaller than the transverse size of the domain occupied by the plasma, that is, $\rho_* \ll X_{max}$, $\rho_* \ll Y_{max}$. Typical values used in the calculations are: $1.5 \leq \rho_* \leq 2.5$, $5 \leq X_{max} = Y_{max} \leq 8$. The longitudinal characteristics of0 the region Ω were chosen from other considerations: $l_* \ll Z_s$, and the value Z_e was determined only by research objectives and available computing resources; the values in the calculations are $l_* = 3.5$, $Z_s = 11$, $Z_e = -12$. Thus, the specified concordance actually means the modelling of the following situation. In front of the pulse ($\eta \geq Z_s$) there is an unperturbed plasma:

$$\psi = -1, \varphi = \gamma = v = 1, \beta = v_x = v_y = v_z = q_x = q_y = b_z = 0.$$

Then (for $Z_e \leq \eta \leq Z_s$), it changes its structure in accordance with the solution of system (9.39)–(9.49), but in the transverse plane the pulse is localized so that the boundary values of the functions (for $|x| = X_{max}$, $|y| = Y_{max}$) can be considered as unperturbed, that is, corresponding to the state of rest. Further, the plasma perturbations lead to a change in the pulse in accordance with equation (9.50), but these changes are concentrated in a certain subdomain that is so far from the Ω boundary that they have practically no effect on the rest state at the periphery. Of course, sooner or later, the perturbations will become significant at the transverse boundaries, but this is completely regulated either by changing the position of the boundaries themselves, or by the time value of the simulation T_{max}.

9.3.2. Difference scheme and solution method

To construct a difference scheme, we will need shifted grids D_i $(0 \leq i \leq 3)$ in the plane of variables (x', y'). Entering the notation $h_x = X_{max}/M_x$, $h_y = Y_{max}/M_y$, where M_x, M_y are the numbers of nodes in the first quadrant with respect to the variables x', y', respectively, we define:

$$D_0 = \{(x_k, y_l): x_k = k h_x, |k| \leq M_x; y_l = l h_y, |l| \leq M_y\},$$
$$D_1 = \{(x_k, y_l): x_k = (k+1/2)h_x, -M_x \leq k \leq M_x - 1; y_l = l h_y, |l| \leq M_y\},$$
$$D_2 = \{(x_k, y_l): x_k = k h_x, |k| \leq M_x; y_l = (l+1/2)h_y, -M_y \leq l \leq M_y - 1\},$$
$$D_3 = \{(x_k, y_l): x_k = (k+1/2)h_x, -M_x \leq k \leq M_x - 1;$$
$$y_l = (l+1/2)h_y, -M_y \leq l \leq M_y - 1\}.$$

Now denoting $h_z = (Z_s - Z_e)/M_z$, where M_z is the number of nodes with respect to the variable η, we construct discrete analogs of the original domain Ω in the form

$$\Omega_{i,h} = \{(x_k, y_l, \eta_j): (x_k, y_l) \in D_i, \ \eta_j = Z_e + j h_z, 0 \leq j \leq M_z\}, \ \ 0 \leq i \leq 3.$$

Different $\Omega_{i,h}$ will be used below as domains of the definition of sets of unknown grid functions: a, ψ, φ, v, vz, β and γ are defined on $\Omega_{0,h}$, q_x and v_x on $\Omega_{1,h}$, q_y and v_y on $\Omega_{2,h}$ and bz on $\Omega_{3,h}$. Hereinafter, the subscripts for functions are used as continuations of names (for example, $b_z \to bz$) to avoid confusion with the indices of grid nodes.

We give difference equations using for the grid functions the notation of the form $f(x_k, y_i, \eta_j, \theta_n) = f_{k,l,j}^n$, $\theta_n = n\tau$, $n \geq 0$, omitting some indices where this should not cause confusion.

First, there is a block of explicit calculations in $\Omega_{0,h}$ for the sequential determination of the functions ψ, vz, v, φ in the direction of decreasing the index j:

$$\frac{\psi_{k,l,j+1} - \psi_{k,l,j}}{h_z} - \beta_{k,l,j+1} = 0, \tag{9.52}$$

$$vz_{k,l,j} + \delta(1 + \psi_{k,l,j}) = 0, \tag{9.53}$$

$$\frac{v_{k,l,j+1}(vz_{k,l,j+1}-1)-v_{k,l,j}(vz_{k,l,j}-1)}{h_z}+$$

$$+\frac{1}{h_x}\left(vx_{k+1,l,j+1}\frac{v_{k+1,l,j+1}+v_{k,l,j+1}}{2}-vx_{k,l,j+1}\frac{v_{k-1,l,j+1}+v_{k,l,j+1}}{2}\right)+ \qquad (9.54)$$

$$+\frac{1}{h_y}\left(vy_{k,l+1,j+1}\frac{v_{k,l+1,j+1}+v_{k,l,j+1}}{2}-vy_{k,l,j+1}\frac{v_{k,l-1,j+1}+v_{k,l,j+1}}{2}\right)=0,$$

$$\varphi_{k,l,j}\psi_{k,l,j}-\Delta\psi_{k,l,j}+v_{k,l,j}(1-vz_{k,l,j})=0. \qquad (9.55)$$

Here it is meant that by the known values on the time layer with the number j + 1 f of the functions ψ, vz, v, φ their values are determined on the time layer with the number j in the internal nodes of the domain D_0. The values at the boundary nodes of the domain D_0 do not change. Thus, integration over the variable η is carried out in the direction of its decrease. The remaining functions involved in the calculations of this block are considered known (previously defined).

The following 7 equations are combined in a block, where the functions β, qx, qy, γ, vx, vy, bz are interconnected in a more complex way, and therefore are defined implicitly:

$$\frac{\beta_{k,l,j+1}-\beta_{k,l,j}}{h_z}-\Delta\gamma_{k,l,j}+\varphi_{k,l,j}\gamma_{k,l,j}-v_{k,l,j}+$$

$$+\frac{1}{h_x h_z}\left(qx_{k+1,l,j+1}-qx_{k+1,l,j}-qx_{k,l,j+1}+qx_{k,l,j}\right)+ \qquad (9.56)$$

$$+\frac{1}{h_y h_z}\left(qy_{k,l+1,j+1}-qy_{k,l+1,j}-qy_{k,l,j+1}+qy_{k,l,j}\right)=0 \quad \text{at } \Omega_{0,h},$$

$$qx_{k,l,j}\frac{\varphi_{k,l,j}+\varphi_{k-1,l,j}}{2}+\frac{\beta_{k,l,j}-\beta_{k-1,l,j}}{h_x}-$$

$$-vx_{k,l,j}\frac{v_{k,l,j}+v_{k-1,l,j}}{2}-\frac{bz_{k,l+1,j}-bz_{k,l,j}}{h_y}=0 \quad \text{at } \Omega_{1,h}, \qquad (9.57)$$

$$qy_{k,l,y}\frac{\varphi_{k,l,j}+\varphi_{k-l-1,j}}{2}+\frac{\beta_{k,l,j}-\beta_{k,l-1,j}}{h_y}-$$

$$-vy_{k,l,j}\frac{v_{k,l,j}+v_{k,l-1,j}}{2}+\frac{bz_{k+1,l,j}-bz_{k,l,j}}{h_x}=0 \quad \text{at } \Omega_{2,h}, \qquad (9.58)$$

$$2\gamma_{k,l,j}\psi_{k,l,j}^2 + \left(\frac{qx_{k,l,j} + qx_{k+1,l,j}}{2}\right)^2 +$$

$$+ \left(\frac{qy_{k,l,j} + qx_{k,l+1,j}}{2}\right)^2 + 1 + \frac{\left|a_{k,l,j}^{n+1}\right|^2}{2} = 0 \quad \text{at} \quad \Omega_{0,h}, \tag{9.59}$$

$$\frac{vx_{k,l,j+1} - vx_{k,l,j}}{h_z} + \delta\frac{qx_{k,l,j+1} - qx_{k,l,j}}{h_z} =$$

$$= \delta\frac{\gamma_{k,l,j} - \gamma_{k-1,l,j}}{h_x} \quad \text{at} \quad \Omega_{1,h}, \tag{9.60}$$

$$\frac{vy_{k,l,j+1} - vy_{k,l,j}}{h_z} + \delta\frac{qy_{k,l,j+1} - qy_{k,l,j}}{h_z} =$$

$$= \delta\frac{\gamma_{k,l,j} - \gamma_{k,l-1,j}}{h_y} \quad \text{at} \quad \Omega_{2,h}, \tag{9.61}$$

$$bz_{k,l,j} - \frac{qx_{k,l,j} - qx_{k,l-1,j}}{h_y} + \frac{qy_{k,l,j} - qy_{k-1,l,j}}{h_x} = 0 \quad \text{at} \quad \Omega_{3,h}. \tag{9.62}$$

Above for the transverse part of the Laplace operator

$$\Delta_\perp = \frac{\partial^2}{\partial(x')^2} + \frac{\partial^2}{\partial(y')^2}$$

in the node (x_k, y_l), the standard five-point approximation 'cross' [22] was used:

$$\Delta f_{k,l} = \frac{f_{k+1,l} - 2f_{k,l} + f_{k-1,l}}{h_x^2} + \frac{f_{k,l+1} - 2f_{k,l} + f_{k,l-1}}{h_y^2}.$$

In the implicit block, as before, the values of variables on the layer $j + 1$ are known, moreover, variables from an explicit block are already calculated. Therefore, equations (9.56)–(9.62) were solved inside the domains D_i ($0 \leq i \leq 3$), using unperturbed values as the boundary values. To find an approximate solution of a system of nonlinear difference equations, the Newton method was used with inversion of the linear part of the operator based on the UMFPACK package [130], designed to solve large systems of equations with asymmetric sparse matrices. The accuracy was fixed maximum in the indices k, l and all equations with a residual value not exceeding $\delta_{\text{newt}} = 10^{-3}$. As a rule, this corresponded to no more than three Newtonian iterations.

Summing up the above, we formulate an iterative procedure for solving equations (9.52)–(9.62). Let at the time θ_{n+1} be known the function $a_{k,l,j}^{n+1}$, which is included in equation (9.59). Then successively for each $j = M_z - 1$, $M_z - 2$, ..., 0, calculations are performed first in an explicit and then in an implicit block, and this process generates the meanings of all functions describing the behavior of the plasma.

Then, from the solution obtained above, the function $\varphi_{k,l,j}$ is taken to complete the equation for the pulse envelope:

$$\frac{2\mathbf{i}}{\varepsilon\theta_*} \frac{a_{k,l,j}^{n+1} - a_{k,l,j}^n}{\tau} + \Delta a_{k,l,j}^{n+1} + \left(1 - \varphi_{k,l,j}\right) a_{k,l,j}^{n+1} = 0.$$

This equation was solved by an iterative method of the type of compressing mappings (s is the iteration number):

$$\frac{2\mathbf{i}}{\varepsilon\theta_*} \frac{a_{k,l,j}^{n+1,s+1} - a_{k,l,j}^n}{\tau} + \Delta a_{k,l,j}^{n+1,s+1} + \left(1 - \varphi_{k,l,j}^s\right) a_{k,l,j}^{n+1,s+1} = 0, \qquad (9.63)$$

where the initial approximation was chosen as $a_{k,l,j}^{n+1,0} = a_{k,l,j}^n$, and $\varphi_{k,l,j}^s$ is a solution of the system (9.52)–(9.62), in which for the function $a_{k,l,j}^{n+1}$ were used its values at the previous iteration, i.e. $a_{k,l,j}^{n+1,s}$. The loop was repeated until

$$\max_{k,l,j} | a_{k,l,j}^{n+1,s+1} - a_{k,l,j}^{n+1,s} | > \delta_{it} = 10^{-3}.$$

After one or two iterations of this type, the desired value of $a_{k,l,j}^{n+1}$ for the moment of time θ_{n+1} was obtained, and the outer loop ended.

Thus, the total time step was a double inserted iterative loop: the inner one was Newtonian for 7 functions of the system (9.56)–(9.62), and the outer one was of the type of compressible mappings for the function a using the relation (9.63).

We make a remark about the orders of approximation of the constructed scheme with respect to the steps in x', y', η, and θ. In terms of the time variable θ, the scheme is chosen completely implicit, which determines the first order approximation; in terms of the spatial variables x' and y', the use of shifted grids leads to a second order approximation. Concerning discretization on the variable η, the situation is more complicated. If we consider the formal approximations of the derivatives included in equations (9.52)–(9.62), then the order of approximation is equal to one. However, the proposed differential scheme has an additional property that manifests itself in problems that are close to linear. We explain this in more detail. Let the perturbations of the background values for the quantities describing the plasma be small, for example,

$$\psi = -1 + \tilde{\psi}, \quad \gamma = 1 + \tilde{\gamma}, \quad |\tilde{\psi}| \ll 1, \quad |\tilde{\gamma}| \ll 1.$$

Then the initial system (9.39)–(9.49) can be linearized in the neighborhood of the background. Omitting the terms of the second order of smallness, we obtain a linear problem, from which, after the elimination of the unknowns, there follows a single basic equation for the potential perturbation:

$$\frac{\partial^2 \tilde{\psi}}{\partial \eta^2} + (1+\delta)\tilde{\psi} + \frac{|a|^2}{4} = 0.$$

The perturbations of the other quantities are expressed explicitly in $\tilde{\psi}$. If such transformations are done with discrete equations (9.52)–(9.62), then the result is the scheme

$$\frac{\tilde{\psi}_{k,l,j+1} - 2\tilde{\psi}_{k,l,j} + \tilde{\psi}_{k,l,j-1}}{h_z^2} + (1+\delta)\tilde{\psi}_{k,l,j} + \frac{|a_{k,l,j}^{n+1}|^2}{4} = 0,$$

obviously having a second order of approximation in the variable η. Test calculations confirm the presence of this property in weakly non-linear versions of the problem.

9.3.3. Calculation results

Numerical simulation was performed for a stationary pulse of the form (9.51) and showed that a short laser pulse ($l_* = 3.5$) effectively excites a wake wave, which, acting on ions, forms a region with a

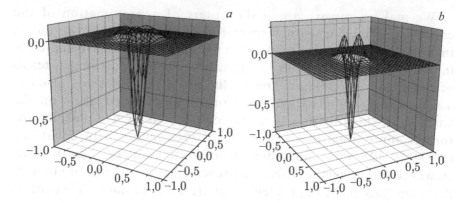

Fig. 9.12. Perturbations of electron density for pulses of circular (*a*) and elliptical (*b*) cross sections after half a period from the centre of the pulse.

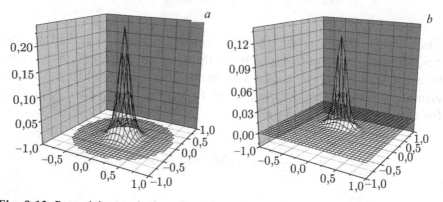

Fig. 9.13. Potential perturbations for pulses of circular (*a*) and elliptic (*b*) sections through a period from the centre of the pulse.

low plasma density (plasma channel) behind the laser pulse. With distance from the pulse the depth of the channel increases, which eventually leads to the destruction of the wake wave.

Of particular interest is the comparison of waves excited by pulses with different cross sections. We present the results of calculations for stationary ions, i.e., at $\delta = 0$, in the domain $\Omega = [-7, 7] \times [-7, 7] \times [-12, 11]$, the centre of the pulse was at the point $\{0, 0, 0\}$, the grid contained $31 \times 31 \times 231$ nodes, the other pulse parameters were $a_* = 1.0$, $\rho_* = 1.8$.

In Fig. 9.12 the electron density perturbations are shown at a distance of half a period from the pulse centre: Fig. 9.12 *a* corresponds to a pulse of circular section ($\alpha = 1$), 9.12 *b* to the elliptical ($\alpha = 4$). The difference in the geometric shape of the functions, arising from the rejection of the axial symmetry of the problem, is clearly noticeable. Note that in this case, the boundaries of change of the considered functions are almost the same.

In Fig. 9.13 potential perturbations are shown through the period from the centre of the pulse and for the same parameters. Here the situation is somewhat different: the geometric shapes are close, but the difference in absolute values is significant (they diverge by more than two times).

The main result of the section can be formulated as follows: to solve a system of nonlinear partial differential equations describing the three-dimensional dynamics of ions and electrons in a plasma wake wave excited by a powerful short laser pulse, a difference scheme and an iterative algorithm for its implementation have been proposed. Starting numerical experiments show that the physical conclusions obtained on the basis of simulating only axially symmetric wake waves can undergo significant changes, since the influence of the geometry of the initiating laser pulse is great not only qualitatively but also quantitatively.

9.4. Bibliography and comments

The axial solution for the wake wave was analyzed in [100]; from its results, in particular, it follows that the wake wave, initiated by an axially symmetric laser pulse, can break over outside the axis of symmetry of the problem. Unfortunately, the author is not aware of other works on the topic of axial solutions.

There is a popular, but very qualified, work on the topic of ion acceleration using laser pulses [35]. Numerical simulation of the dynamics of ions in three-dimensional nonlinear axially symmetric wake waves in the framework of the hydrodynamic model was apparently carried out for the first time in [125]. Subsequently, the numerical algorithm developed there was used for calculations in [136]. Within the framework of the PiC-simulation, the dynamics of ions began to be studied, starting, apparently, with the work [159].

The rejection of axial symmetry when taking into account the motion of ions in a three-dimensional wake wave greatly complicates the formulation of the problem. A numerical simulation method based on the hydrodynamic model was apparently first considered in [48]. Analogues of such calculations, unfortunately, are not found in the literature by the author.

The category of additional studies on the breaking of wake waves in the hydrodynamic approximation is the work [93]. It carried out a numerical and analytical study of the spatial structure of the wake wave excited in an underdense plasma by a focused laser pulse. In a strongly nonlinear mode, calculations of the spatial distribution of density and electric fields in the wake wave were carried out using the finite difference method and the particle method. It is shown numerically that, when moving away from the back front of a laser pulse, a gradual formation of a maximum electron density occurs, located at a certain distance from the axis. After this off-axis density maximum is formed, its value increases rapidly from period to period. At a certain distance from the back front of the pulse, which depends essentially on the amplitude of the laser field, the value of the off-axis density maximum becomes infinitely large and the wake wave breaks. Numerical calculations by the particle method show that the density singularity arises due to the intersection of the trajectories of neighbouring electrons. An increase in the amplitude of the laser field leads to a rapid approach of the breaking point to the trailing edge of the pulse, with the result that the breaking can occur even inside the body of the laser pulse. In the weak nonlinearity mode, an expression for the nonlinear frequency shift of the radial oscillations of electrons in the wake wave is derived using the perturbation method. An analytical dependence of the breaking point of a weakly nonlinear wake wave on the amplitude of the laser field is found, which coincides with the accuracy of a numerical coefficient with the results of calculations by the particle method in the case of strong nonlinearity. The analysis was carried out in the approximation of a given pulse. The conclusion that the wake wave, initiated by an axially symmetric laser pulse, can break over off-axis, was made, inter alia, on the basis of studies of this work.

Conclusion

The preface addressed the topic of orientation of the material of the book into the future, i.e., attracting the attention of researchers to new problem statements. We note here several possible directions.

Crashing solutions of quadratic systems of ordinary differential equations. Issues related to the existence of global solutions appear almost always when differential equations are considered. Here are two of them that have not been fully studied to date [17]:

1. When is the interval bounded or unlimited, in which some solutions of a system of differential equations exist?

2. How to effectively find this interval?

The book made the first attempts to address these issues for problems related to axial solutions. It seems promising to study the statements associated with spherically symmetric vibrations, as well as taking into account the influence of the dynamics of ions on the electronic oscillations with different symmetries of the initial problem.

Diagnostics of gradient catastrophe for solving a quasilinear system of hyperbolic equations. If we consider the Cauchy problem for such a system in invariants [81], then with limited right-hand sides, the solutions remain bounded (in absolute value), but their derivatives can nevertheless increase without limit. This effect is called gradient catastrophe. In the case of plasma oscillations, the occurrence of such a situation means the limit of applicability of the hydrodynamic model used, since it leads to an infinite concentration of charge. The problem is complicated by the fact that a gradient catastrophe can occur after a certain (sometimes very significant) number of periods of oscillation. An attempt to study solutions of this kind related to plasma oscillations was made in [104], but the results obtained in it should be considered only as a primary acquaintance with the formulation of the problem.

Constructing numerical algorithms for rollover problems. In this direction, methods in Euler variables, which are of the second order of accuracy up to tipping, are of natural interest. The special features of the formulation are the non-divergent form of the equations and the fundamental absence of the auxiliary Riemann problem, that is, problems with piecewise constant (or, in the general case, piecewise linear) initial data. The method should also have a shock absorbing property, i.e. Be able to track the resulting discontinuity of functions and / or their derivatives, but in the absence of any information about its location and time, as well as about the asymptotics of the desired quantities in the vicinity of the discontinuity.

Of particular interest are algorithms designed to take into account the dynamics of ions, since in this case the calculated number of oscillation periods can increase by tens (or even hundreds!) Times. Preserving the accuracy of calculations over long times is a known difficulty in modeling quasiperiodic solutions.

The study of the effect of overturning plasma oscillations in the framework of other mathematical models. Of primary interest are integro-differential (kinetic) equations of the Vlasov – Poisson or Vlasov – Maxwell type. For one-dimensional plane oscillations, to formulate the corresponding Cauchy problem for the Vlasov equations, both in the non-relativistic case and taking into account the relativistic effect, is not particularly difficult. Moreover, for the nonrelativistic formulation, the classical theorem on the existence of a solution is known [61]. However, when you try to numerical simulation immediately occurs the question of artificial boundary conditions, first of all, in the direction of change in the independent spatial variable. In the review [132], when formulating spatially one-dimensional test problems, here, as a rule, periodic boundary conditions are used. The construction and analysis of other types of conditions is put into an independent problem.

Also of interest is a formal comparison of numerical experiments on hydrodynamic models and particle-in-cell models. We note once again the fundamental difficulty of using PiC-models for modeling spatially multi-dimensional plasma oscillations. This is due to the assignment of the initial conditions, first of all, to the initial distribution of particles. The fact is that in the hydrodynamic model of oscillations at the initial moment of time, a function is described that describes the electric field. In the methods of particles, they should be arranged at the initial moment of time (shifted relative to the equilibrium position) so that the displacement of the particles just formed a given field. This leads to the need to solve in some sense the inverse problem, which is also an independent problem.

The presence of promising areas is an important source for the development of science in the subject area. However, we emphasize that the proposed book is devoted to mathematical aspects, that is, to particular views from different points of view on the properties of equations, solutions, methods, etc., arising from the simulation of oscillations and wake waves in a plasma. And there is no reason to believe that other (perhaps alternative) views are less significant in the study of similar problems.

References

1. SKIF MSU Chebyshev. https://parallel.ru/claster/superinfo.
2. SPECFUN. http://www.netlib.no/netlib/specfun.
3. *Abdullaev A.Sh., Aliev Yu.M., Frolov A.A.*, Fizika plazmy. 1986. V. 12, No. 7. P. 827–835.
4. *Abdullaev A.Sh., Frolov A.A.*, Zh. Eksper. Teor. Fiz. 1981. V. 81, No. 3 (9). P. 927–932.
5. *Agoshkov V.I., Dubovskiy P.B., Shutyaev V.P.*, Methods for solving problems of mathematical physics. Moscow, FIZMATLIT, 2002.
6. *Aleksandrov A.F., Bogdankevich L.S., Rukhadze A.A.* Basics of plasma electrodynamics. Ed. 2nd, Moscow, Vysshaya shkola, 1988.
7. *Anderson D., Tannehil J., Pletcher R.* Computational fluid mechanics and heat transfer. V. 1. Moscow, Mir, 1990.
8. *Andreev N.E., Gorbunov L.M., Zykov A.I., Chizhonkov E.V.*, Zh. Eksper. Teor. Fiz. 1994. V. 106, No. 6. P. 1676–1686.
9. *Andreev N.E., Gorbunov L.M., Kirsanov V.I., Pogosova A.A., Ramazashvili R.R.* Pis'ma Zh. Eksper. Teor. Fiz. 1992. V. 55, No. 10. P. 551–555.
10. *Andreev N.E., Gorbunov L.M., Ramazashvili R.R.*, Fizika plasmy. 1997. V. 23, No. 4. P. 303–310.
11. *Andreev N.E., Gorbunov L.M., Frolov A.A., Chizhonkov E.V.*, in: Physics of Extreme States of Matter – 2005 / Ed. V.E. Fortov and others. Chernogolovka: IPCP RAS. 2005. P. 12–14.
12. *Andreev N.E., Gorbunov L.M., Chizhonkov E.V.*, Matem. Modelirovanie. 1995. V. 7, No. 9. No. 65–71.
13. *Arnold V.I.*, Ordinary differential equations. Moscow, Nauka, 1971.
14. *Arnold V.I.*, Theory of catastrophes. Itogi nauki tekhniki. Ser. Sovr. probl. mat. Fund. napravleniya. 1986. V. 5. P. 219–277.
15. *Akhiezer A.I., Polovin R.V.*, Zh. Eksper. Teor. Fiz. 1956. V. 30, No. 5. P. 915–928.
16. *Babenko K.I.*, Basics of numerical analysis. Moscow, Nauka, 1986.
17. *Baris Ya., Baris P., Ruhlevich B.* Disruptive solutions of quadratic systems of differential equations. Sovremen. Matematika. Fundament. napravleniya. 2006. V. 15. P. 29–35.
18. *Bakhvalov N.S., Borovskiy A.V., Korobkin V.V., Chizhonkov E.V., Eglit M.E., Yakubenko A.E.*, Numerical calculation of the gains of the light on transitions of H-ions in a freely expanding plasma. Preprint IOF Academy of Sciences of the USSR. 1985. No. 187. P. 34.
19. *Bakhvalov N.S., Borovskiy A.V., Korobkin V.V., Chizhonkov E.V., Eglit M.E.*, Heating and non-equilibrium thermal ionization of a plasma by a short laser pulse. Preprint IOF, USSR Academy of Sciences. 1987.No. 166. P. 21.

20. *Bakhvalov N.S., Zhidkov N.P., Kobelkov G.M.* Numerical methods. 7th ed. Moscow, BINOM. Laboratory of knowledge, 2013.
21. *Bakhvalov N.S., Zhilekin Ya.M., Zabolotskaya E.A.* Nonlinear theory of sound beams. Moscow, Nauka, 1982.
22. *Bakhvalov N.S., Kornev A.A., Chizhonkov E.V.,* Numerical methods. Problem solving and exercise. 2nd ed. Moscow, Laboratory of knowledge, 2016.
23. *Belotserkovsky OM, Davydov Yu.M.* The method of large particles in gas dynamics. Moscow, Nauka, 1982.
24. *Bogachev K.Yu.* Basics of parallel programming. Moscow, BINOM. Laboratory of knowledge, 2013.
25. *Bogolyubov N.N., Mitropolsky Yu.A.* Asymptotic methods in the theory of nonlinear oscillations. Moscow, Nauka, 1974.
26. *Borisov A.B.* Symmetric Crank–Nicholson type schemes for solving nonlinear Schrödinger equations. In: Methods and algorithms for numerical analysis and their applications. Moscow, MGU Publishing House, 1989. P. 153–181.
27. *Borovskiy A.V., Galkin A.L., Priymak V.G., Chizhonkov E.V.,* Zh. Vychisl. Mat. i Mat. Fiziki. 1990. V. 30, No. 9. P. 1381–393.
28. *Borovskiy A.V., Staroverov V.M., Chizhonkov E.V.,* Influence of ionization-recombination processes going through the excited states of ions on the evolution of a multiply charged plasma in gas-dynamic calculations. Preprint IOF, USSR Academy of Sciences. 1989. No. 32. P. 19.
29. *Borovskiy A.V., Galkin, A.L.,* Laser physics: X-ray lasers, ultrashort pulses, powerful laser systems. Moscow, IzdAT, 1996.
30. *S. Braginsky.* Transport phenomena in plasma. In: Questions of the theory of plasma. Gosatomizdat, Moscow, 1963. P. 183–285.
31. *Brushlinsky K.V.,* Mathematical and computational problems of magnetic gas dynamics. Moscow, BINOM. Laboratory of Knowledge, 2009.
32. *Bulanov, SV, Esirkepov, T.Zh., Kando, M., et al.,* Uspekhi Fiz. Nauk. 2013. V. 183, No. 5. P. 449–486.
33. *Bulanov S.V., et al.,* Fizika plazmy.. 1999. V. 25, No. 6. P. 517–530.
34. *Bulanov S.V., et al., ibid,* 2006. V. 32, No. 4. P. 291–310.
35. *Bychenkov V.Yu.,* Priroda.. 2012. No. 2. P. 3–11.
36. *Bychenkov Yu.V., Chizhonkov E.V.,* Iterative methods for solving saddle problems. Moscow, BINOM. Laboratory of Knowledge, 2010.
37. *Vatazhin A.B., Lyubimov G.A., Regirer S.A.,* Magnetohydrodynamic flows in the channels. Moscow, Nauka, 1970.
38. *Vedenyapin V.V.,* Kinetic equations of Boltzmann and Vlasov. Moscow, FIZMATLIT, 2001.
39. *Gelfand I.M., Zueva N.M., Imshennik V.S., Lokutsievsky O.V., Ryabenky V.S., Khazina L.G.* Zh. Vychisl. Mat i Mat. Fiziki.. 1967. V. 7, No. 2. P. 322–347.
40. *Ginzburg V.L., Rukhadze A.A.,* Waves in a magnetoactive plasma. Moscow, Nauka, 1975.
41. *Godunov S.K., Ryabenky V.S.,* Difference schemes. Introduction to the theory. Moscow, Nauka, 1973.
42. *Gorbunov L.M.,* Usp. Fiz. Nauk. 1973. V. 109, No. 4. P. 631–655.
43. *Gorbunov L.M.,* Priroda.. 2007, No. 4. P. 11–20.
44. *Gorbunov L.M., Kirsanov V.I.* Zh. Eksper. Teor. Fiz. 1987. Vol. 93, No. 2. P. 509–518.
45. *Gorbunov L.M., Frolov A.A., Chizhonkov E.V.,* On modeling nonrelativistic cylindrical oscillations in a plasma // Vychisl. Metody i Programm. 2008. V. 9, No. 1. P. 58–65.

46.   *Gorbunov L.M., Frolov A.A., Chizhonkov E.V., Andreev N.E.,* Fizika plazmy, 2010. V. 36, No. 4. P. 375–386.
47.   *Gorbunov L.M., Chizhonkov E.V.,* Fundamental'naya i prikladnaya matematika. 1996. V. 2, No. 3. P. 789–801.
48.   *Gorbunov L.M., Chizhonkov E.V.,* Vychisl. metody i programm., 2006. V. 7. P. 17–22.
49.   *Gorbunov L.M., Chizhonkov E.V.,* in: Analytical and numerical methods for modeling natural science and social problems: Coll. Articles II Intern. scientific and technical conference. Penza, ANOO Volga Knowledge House. 2007. P. 277–279.
50.   *Grigoriev Yu.N., Vshivkov V.A., Fedoruk M.P.,* Numerical simulation with particle-in-cell methods. Novosibirsk: Publishing House of the Siberian Branch of the Russian Academy of Sciences, 2004.
51.   *Dwight G.B.,* Integral tables and other mathematical formulas. Moscow, Nauka, 1973.
52.   *Jackson D.,* Fourier series and orthogonal polynomials. Moscow, GITTL, 1948.
53.   *Dniester Yu.N., Kostomarov D.P.,* Mathematical modeling of plasma. Moscow, Nauka, 1982.
54.   *Evstigneev V.A.,* NUMA-architecture: some features of compilation and code generation, in: Support for supercomputers and Internet-oriented technologies. Novosibirsk: ISI SB RAS, 2001.
55.   *Elizarova T.G.,* Quasi-gasdynamic equations and methods for calculating viscous flows. Moscow, Nauchnyi mir, 2007.
56.   *Elizarova T.G., Chetverushkin B.N.,* Zh. Vychisl. Mat. Mat. Fiziki. 1985. V. 25, No. 10. P. 1526–1533.
57.   *Zakharov V.E., Synakh V.S.,* Zh. Eksper. Teor. Fiz., 1975. V. 68, No. 3. P. 940–947.
58.   *Zeldovich Ya.B., Mamaev A.V., Shandarin S.F.,* Uspekhi Fiz. Nauk, 1983. V. 139, No. 1. P. 153–163.
59.   *Zeldovich Ya.B., Myshkis A.D..* Elements of mathematical physics. Moscow, Nauka, 1973.
60.   *Ilgamov M.A., Gilmanov A.N.,* Non-reflective conditions at the boundaries of the computational domain. Moscow, FIZMATLIT, 2003.
61.   *Iordansky S.V.,* Trudy MIAN SSSR. 1961. V. 60. P. 181–194.
62.   *Kamke E.,* Handbook of differential equations. Moscow, Nauka, 1965.
63.   *Karamzin Yu.N.,* Difference methods in problems of nonlinear optics. Moscow, Preprint IPM im M.V. Keldysha, Academy of Sciences of the USSR. No. 74, 1982.
64.   *Karchevsky M.M., Pavlova M.F.,* Equations of mathematical physics. Additional chapters. St Petersburg, Publishing house Lan, 2016.
65.   *Katok A.B., Hasselblat B.,* Introduction to the modern theory of dynamical systems. Moscow, Faktorial Publishing House, 1999.
66.   *Coddington E.L., Lewinson N.,* Theory of ordinary differential equations. Moscow, Publishing House of Foreign Literature, 1958.
67.   *Konik A.A., Chizhonkov E.V.,* Difference method for modeling wake waves in a plasma, in: Materials of the Tenth Intern. conf. Kazan: Publishing House of Kazan State. University, 2014. P. 391–397.
68.   *Konik A.A., Chizhonkov E.V.,* Moscow Univ. Bulletin. Series 1. Mathematics, mechanics. 2016, No. 1. P. 44–48.
69.   *Kulikovskiy A.G., Pogorelov N.V., Semenov F.Yu.,* Mathematical problems of numerical solution of hyperbolic systems of equations. Moscow, FIZMATLIT, 2001.
70.   *Lebedev V.I.,* Zh. Vychisl. Mat Mat. Fiziki. 1964. Vol. 4, No. 3. P. 449–465.
71.   *Lebedev V.I.,* Functional analysis and computational mathematics. Moscow, FIZMATLIT, 2005.

72. *Leng S.* Elliptic functions. Moscow, Nauka, 1984.

73. *Lugovoy V.N., Prokhorov A.M.,* Usp. Fiz. Nauk, 1973. V. 111, No. 2. P. 203–247.

74. *Marchuk G.I.* Methods of computational mathematics. Moscow, Nauka, 1980.

75. *Milyutin S.V., Frolov A.A., Chizhonkov E.V.,* in: Supercomputer technologies in science, education and industry. V. 4 / Ed. Academician V.A. Sadovnichiy, et al., Moscow, Publishing House of Moscow University, 2012.

76. *Milyutin S.V., Frolov A.A., Chizhonkov E.V.,* Vychisl. metody i programm. 2013. V. 14, No. 2. C. 295–05.

77. *Milyutin S.V., Frolov A.A., Chizhonkov E.V.,* in: Proc. Scientific International conf. Difference schemes and their applications dedicated to the 90th anniversary of Professor V.S. Ryaben'ky. Moscow: Inst. Of Applied Mathematics of M.V. Keldysh, RAS, 2013. P. 105–106.

78. *Morozov A.I., Soloviev L.S.* Stationary plasma flows in a magnet field. /Questions of the theory of plasma. Moscow, Atomizdat, 1974.

79. *Popov A.V., Chizhonkov E.V.,* Vychisl. metody i programm. 2012. V. 13, No. 1. P. 5–17.

80. *Pokhozhaev S.I.,* Tr. Matem Inst. V.A. Steklova. 2003. V. 243. P. 257–288.

81. *Christmas B.L., Yanenko N.N.,* Systems of quasilinear equations and their application to gas dynamics. Moscow, Nauka, 1968.

82. *Samarskii A.A.,* Introduction to the theory of difference schemes. Moscow, Nauka, 1971.

83. *Samarskii A.A., Nikolaev E.S.,* Methods for solving grid equations. Moscow, Nauka, 1978.

84. *Samarskii A.A.,* Sobol' I.M., Zh. Vychisl. Matem. Matem. Fiziki. 1963. Vol. 3, No. 4. P. 702–719.

85. *Silin V.P.,* Introduction to the kinetic theory of gases. Moscow, Nauka, 1971.

86. *Silin V.P., Rukhadze A.A.,* Electromagnetic properties of plasma and plasma-like media. MoscowTrading House Librokom, 2012.

87. *Sobolev S.L.,* Equations of mathematical physics. 4th ed. Moscow, Nauka, 1966.

88. Tikhonov A.N., Samarskii A.A., Equations of mathematical physics. Moscow, Nauka, 1972.

89. *Tyrtyshnikov E.E.,* Numerical analysis methods. Mocow, Akademiya, 2007.

90. *Fedorenko R.P.,* Introduction to computational physics. Moscow, Publishing House of the Moscow Institute of Physics and Technology, 1994.

91. *Fedorova I.V., Chizhonkov E.V.,* in: Mathematical ideas of P.L. Chebyshev and their application to modern problems of natural science: Abstracts of the Intern. conf. 2008. P. 82–83.

92. *Fedorova I.V., Chizhonkov E.V.,* Vestn. Mosk. un-ta. Ser. 1, Mathematics. Mechanics. 2009, No. 5. P. 50–53.

93. *Frolov A.A., Chizhonkov E.V.,* Fizika plazmy. 2011. V. 37, No. 8. P. 711–728.

94. *Frolov A.A., Chizhonkov E.V.,* Vychisl. metody i programm. 2014. V. 15. P. 537–548.

95. *Frolov A.A., Chizhonkov E.V.,* Matem. modelirovanie, 2015. V. 27, No. 12. P. 3–19.

96. *Frolov, A.A., Chizhonkov, E.V.,* Zh. Vychisl. Matem. Matem. Fiziki. 2017. V. 57, No. 11. P. 1844–1859.

97. *Chetverushkin* B.N., Kinetic schemes and quasi-gas-dynamic systems of equations. Moscow, MAKS Press, 2004.

98. *Chizhonkov E.V.,* Dokl. AN SSSR, 1984. V. 278, No. 5. P. 1074–1077.

99. *Chizhonkov E.V.,* Relaxation methods for solving saddle problems. Moscow, IVM RAS, 2002.

100. *Chizhonkov E.V.,* Vychisl. metody i programm. 2010. V. 11, No. 2. P. 57–69.

101. *Chizhonkov E.V.,* in: Grid methods for boundary value problems and applications. Materials of the Eighth All-Russian Conf., dedicated to the 80th anniversary of A.D. Lyashko. Kazan: Publishing House of Kazan State Univ.. 2010. P. 474–482.

102. *Chizhonkov E.V.,* Zh. Vychisl. Matem. Matem. Fiziki, 2011. V. 51, No. 3. P. 456–469.

103. *Chizhonkov E.V.,* in: Grid methods for boundary value problems and applications. Materials of the Eleventh Intern. conf. Kazan: Publishing House of Kazan State un-Univ., 2016. P. 321–325.

104. *Chizhonkov E.V.,* Vychisl. metody i programm. 2017. V. 18. P. 65–79.

105. *Shen I.R.,* Principles of nonlinear optics. Moscow, Nauka, 1989.

106. *Aliev Yu.M., Stenflo L.,* Physica Scripta. 1994. V. 50. P. 701–702.

107. *Amiranashvili Sh., Yu M.Y., Stenflo L., Brodin G., Servin M.,* Physical Review. 2002. V. 66. P. 046403–1–046403–6.

108. *Andreev N.E., Chizhonkov E.V., Frolov A.A., Gorbunov L.M.,* Nucl. Instr. Meth. in Phys. Res. Sect. A. 1998. V. 410. P. 469–476.

109. *Andreev N.E.. Chizhonkov, E.V., Gorbunov L.M.,* Rus. J. Numer. Anal. Math. Modeling. 1998. V. 13, No. 1. P. 1–11.

110. *Andreev N.E., Chizhonkov E.V., Gorbunov L.M.,* Laser Optics'98: Superstrong Laser Fields and Applications / Ed. by A.A.Andreev. Proc. of SPIE. 1998. V. 3683. P. 2–8.

111. *Andreev N.E., Chizhonkov E.V., Gorbunov L.M., Ramasashvili R.R.,* Structure of 3-D Nonlinear Plasma Waves Generated by Intense Laser Pulse. Proc. of the Intern. Conf. on Lasers-97. STS Press. McLean, Virginia, 1998. P. 831–838.

112. *Andreev N.E., Frolov A.A., Kuznetsov S.V., Chizhonkov E.V., Gorbunov L.M.,* in: Proc. of the Intern. Conf. on Lasers-97. STS Press. McLean. Virginia, 1998. P. 875–881.

113. *Andreev N.E., Gorbunov L.M., Kirsanov V.I., Nakajima K., Odata A.,* Phys. Plasmas. 1997. V. 4. P. 1423–1432.

114. *Andreev N.E., Gorbunov L.M., Tarakanov S.V., Zykov A.I.,* Phys. Fluids. 1993. V. B5. P. 1986–1999.

115. *Birdsall C.K., Langdon A.B.,* Plasma Physics via Computer Simulation. New York: McGraw-Hill Inc., 1985.

116. *Borovskii A.V., Chizhonkov E.V., Galkin A.L., Korobkin V.V.,* Appl. Phys. 1990. V. B, No. 50. P. 297–302.

117. *Bulanov S.V., Maksimchuk A., Schrotder C. B., Zhidkov A.G., Esarey E., Leemans W.P.,* Phys. of Plasma. 2012. V. 19. P. 020702 (1–4).

118. *Bulanov S.V., Pegoraro F., Pukhov A.M.,* Phys. Rev. Lett. 1995. V. 74, No. 5. P. 710–713.

119. *Bulanov S.V., Pegoraro F., Pukhov A.M., Sakharov A.S.,* Phys. Rev. Lett. 1997. V. 78, No. 22. P. 4205–4208.

120. *Chandrasekhar S.,* Ellipsoidal Figures of Equilibrium. New Haven: Yale Univ. Press, 1969.

121. *Chizhonkov E.V.,* Rus. J. Numer. Anal. Math Modeling. 2017. V. 32, No. 1. P. 13–26.

122. *Chizhonkov E.V., Frolov A.A.,* Rus. J. Numer. Anal. Math. Modeling. 2011. V. 26, No. 4. P. 379–396.

123. *Chizhonkov E.V., Frolov A.A., Gorbunov L.M.,* Rus. J. Numer. Anal. Math. Modeling. 2008. V. 23, No. 5. P. 455–467.

124. *Chizhonkov E.V., Frolov A.A., Milyutin S.V.,* Rus. J. Numer. Anal. Math. Modeling. 2015. V. 30, No. 4. P. 213–226.

125. *Chizhonkov E.V., Gorbunov L.M.,* Rus. J. Numer. Anal. Math Modeling. 2001. V. 16, No. 3. P. 235–246.

126. *Chizhonkov E.V., Gorbunov L.M.,* Rus. J. Numer. Anal. Math Modeling. 2007. V. 22, No. 6. P. 531–541.

127. *Cohen B.I., Lasinski B.F., Langdon A.B., Cummings J.C.,* Phys. Fluids. 1991. V. B3, No. 3. P. 766–775.

128. *Cowell W.R.* (editor), Sources and Development of Mathematical Software. Englewood Cliffs. New Jersey: Prentice-Hall Inc., 1984.

129. *Davidson R.C.,* Methods in Nonlinear Plasma. New York: Academic Press, 1972.

130. *Davis T.A.,* UMFPACK Version 4.3 User Guide. Technical Report TR-04-003. Univ. of Florida, CISE Dept., Gainesvill, FL, 2004.

131. *Dawson J.M.,* Phys. Rev. 1959. V. 113, No. 2. P. 383–387.

132. *Dimarco G., Pareschi L.,* Acta Numerica. 2014. V. 23. P. 369–520.

133. *Esarey E., Sprangle P., Krall J., Ting A.,* IEEE Trans. on Plasma Science. 1996. V. 24. P. 252–288.

134. *Fonseca R.A., et al.,* Lecture Notes in Computer Science. 2002. V. 2331. P. 342–351.

135. *Fuchs J., et al.,* Phys. Rev. Lett. 1998. V. 80, No. 8. P. 1658–1661.

136. *Gorbunov L.M., Mora P., Solodov A.A.,* Phys. Rev. Lett. 2001. V. 86, No. 15. P. 3332–3335.

137. *Goriely A., Hyde C.,* Differential Equations, 2000. V. 161. P. 422–448.

138. *Hairer E., Wanner G.* Solving Ordinary Differential Equations II. Stiff and Differential-Algebraic Problems. Second revised ed. Springer Verlag, 1996.

139. *Hockney R.W., Eastwood J.W.,* Computer Simulation Using Particles. New York: McGraw-Hill Inc., 1981.

140. *Huang C., et al.,* J. Physics: Conference Series. 2006. V. 46. P. 190–199.

141. *Infeld E., Rowlands G., Skorupski A.A.,* Phys. Rev. Lett. 2009. V. 102. P. 145005 (1–4).

142. *Jacobson D.H.* Extensions of Linear – Quadratic Control, Optimization and Matrix Theory. London: Academic Press, 1977.

143. *Karimov, A.R., Yu. M.Y., Stenflo L.,* Phys. Plasmas. 2012. V. 19. P. 092118 (1–5).

144. *Kim J.K., Umstadter D.* in: Advanced Accelerator Concepts: Eighth Workshop / Ed. by W. Lawson, C. Bellamy, D. Brosius. AIP Conf. Proc. 472. New York: AIP Press, 1999. p. 404–412.

145. *Kosinski W.,* J. of Mathem. Analysis and Applications., 1977. V. 61. P. 672–688.

146. *Kroll J., Esarey, E., Sprangle, P., Joice G.,* Phys. Plasmas. 1994. V. 1. P. 1738–1743.

147. *Krushelnick K., et al.,* Phys. Rev. Lett. 1999. V. 83, No. 4. P. 737–740.

148. *Krushelnick, K., et al.,* Rev. Lett. 1997. V. 78, No. 21. P. 4047–4050.

149. *Lehmann G., Laedke E.W., Spatschek K.H.,* Phys. Plasmas. 2007. V. 14. P. 103109 (1–9).

150. *MacLeod A.J.,* J. of Comput. and Appl. Math 2002. V. 145, No. 1. P. 237–246.

151. *Machalinska-Murawska J., Szydlowski M.,* Archives of Hydro-Engineering and Environmental Mechanics. 2013. V. 60, No. 1-4. P. 51–62.

152. *Max C.E.,* Phys. Fluids. 1976. V. 19, No. 1. P. 74–77.

153. *Mora P. A.,* Phys. Plasmas. 1997. V. 4, No. 1. P. 217–229.

154. *Nayfeh A.H.,* Introduction to Perturbation Techniques. New York: Jon Wiley and Sons, 1981.

155. *Nieter C., Cary J.B.,* J. Comput. Phys. 2004. V. 196. P. 448–473.

156. *Piessens R., deDoncker-Kapenga E., Uberhuber C., Kahaner D.* Quadpack: a Subroutine Package for Automatic Integration. Springer Verlag, Series in Computational Mathem. V. 1, 1983.

157. *Pohozaev S.I.,* Theory Blow-Up Theory for Nonlinear PDE's. Function Spaces, Differential Operators and Nonlinear Analysis. The Hans Triebel Anniversary Volume. 2003. P. 141–159.

158. *Pukhov A.,* J. Plasma Phys. 2001. V. 61. P. 425–433.

159. *Pukhov A.*, Phys. Rev. Lett. 2001. V. 86, No. 16. P. 3562–3565.
160. *Rosenbluth M.N., Liu C.S.*, Phys. Rev. Lett. 1972. V. 29, No. 11. P. 701–705.
161. *Rowlands G., Brodin G., Stenflo L.*, J. Plasma Phys. 2008. V. 74, No. 4. P. 569–573.
162. *Sprangle P., Esarey E., Ting A., Joyce G.*, Appl. Phys. Lett. 1988. V. 53. P. 2146–2148.
163. *Stenflo L.*, Phys. Scripta 1996. V. 63. P. 59–62.
164. *Stenflo L., Gradov O.M.*, Phys. Rev. E. 1998. V. 58, No. 6. P. 8044–8045.
165. *Stenflo L., Marklund M., Brodin G., Shukla P.K.*, J. Plasma Phys. 2006. V. 72, No. 4. P. 429–433.
166. *Tajima T., Dawson J.M.*, Phys. Rev. Lett. 1979. V. 43. P. 267–270.
167. *Verboncoeur J.P.*, Plasma Phys. and Controlled Fusion. 2005. V. 47. P.A231 – A260.
168. *Verboncoeur J.P., Langdon A.B., Gladd N.T.*, Comput. Phys. Commun. 1995. V. 87 (1–2). P. 199–211.
169. *Verma P.S., Soni J.K., Segupta S., Kaw P.K.*, Phys. Plasmas. 2010. V. 17. P. 044503 (1–4).
170. *Yee K.S.*, IEEE Trans. 1996. V. 14. P. 302–307.
171. *Young P.E., Guethlein G., Wilks S.C., Hammer J.H., Kruer W.L., Stabrook K.G.*, Phys. Rev. Lett. 1996. V. 76, No. 17. P. 3128–3131.

Index

A

algorithm
  tridiagonal algorithm 221, 228
approximation
  Galerkin approximations 242
  quasistatic approximation 186, 189, 271

B

breaking 1, 2, 4, 5, 9, 11, 12, 14, 16, 17, 18, 19, 22, 49, 51, 52, 56, 60,
    61, 62, 63, 66, 67, 68, 69, 70, 71, 72, 73, 74, 75, 76, 133, 137,
    141, 144, 145, 146, 147, 148, 149, 150, 152, 154, 157, 158, 164,
    165, 169, 170, 171, 172, 173, 174, 177, 178, 179, 180, 181, 182,
    184, 193, 205, 224, 234, 237, 239, 241, 242, 243, 253, 254, 255,
    268, 269, 281

E

envelope 13, 111, 113, 184, 187, 198, 205, 207, 208, 223, 258, 270, 273,
    277
equation
  P1NE equations 23, 50
  Maxwell's equations 5, 7, 186
  P1RE equations 49, 50

F

factor
  Lorentz factor 113, 152, 160, 186, 261

frequency
  plasma frequency 10, 78, 134, 153, 184, 190, 258, 270

G

gradient catastrophe 19, 60, 75, 76, 150, 173, 239, 282

I

inequality
  Gronwall inequality 9, 21

L

law of conservation of a generalized curl 205

M

method
  finite difference method 107, 113, 130, 131, 133, 222, 229, 236, 237, 242, 281
  linearization method 237, 243
  matrix Thomas (diagonal) method 216
  projection method 230, 233, 234, 236, 237
  projection (spectral) method 229, 237, 242
  Runge–Kutta method 74, 234
model
  numerical–asymptotic model 152
  PiC-models 182, 283

P

P1NE (Plane 1-dimension Nonrelativistic Electron oscillations) 11
P1RE (Plane 1-dimension Relativistic Electron oscillations) 11
problem
  Cauchy problem 6, 7, 15, 19, 26, 28, 30, 31, 66, 73, 75, 83, 84, 86, 88, 124, 126, 142, 143, 195, 197, 199, 201, 202, 226, 263, 282, 283
  'complete' problem 252, 253, 255, 258
  'truncated' problem 247, 251, 254, 255
pulse
  'slow' pulse 202, 203, 204, 205

S

scheme
  Lax–Wendroff 63, 73, 91, 138
  leapfrog' scheme 49, 57, 252
  McCormack 73
  tripod' scheme 64, 92, 139
solution
  axial solution 25, 35, 83, 106, 110, 123, 141, 145, 247, 248, 255, 280
  polygonal' solution 48
  triangular solution 36, 40
system
  P1NE system 14, 22, 23, 47

P1RE system 49

T

theorem
  Poincaré – Bendixson theorem 29, 46. 86

theory
  Poincaré–Bendixson theory 45
time
  breaking time 19, 56, 61, 62, 63, 145, 158, 177
transverse breaking 241, 242

W

wake wave 130, 149, 154, 184, 185, 187, 188, 193, 196, 204, 205, 208,
      209, 213, 215, 217, 220, 223, 224, 229, 234, 236, 241, 242, 243,
      245, 246, 247, 253, 254, 258, 267, 268, 269, 278, 279, 280, 281